1986

Hazardous and
Toxic Materials

HAZARDOUS AND TOXIC MATERIALS
Safe Handling and Disposal

Howard H. Fawcett, P.E.
President, Fawcett Consultations, Inc.

A Wiley-Interscience Publication
JOHN WILEY & SONS
New York   Chichester   Brisbane   Toronto   Singapore

Library of Congress Cataloging in Publication Data:

Fawcett, Howard H.
  Hazardous and toxic materials.

  "A Wiley-Interscience publication."
  Bibliography: p.
  Includes index.
  1. Hazardous wastes—United States—Safety measures.
2. Poisons—Safety measures. I. Title.
TD811.5.F39 1984   363.1'79   84-5148
ISBN 0-471-80483-5

Printed in the United States of America

10 9 8 7 6 5 4 3 2

To
Ruth, Ralph Willard, and Harry Allen,
my family
without whose enthusiastic support
this volume would not exist.

Preface

When any aspect of science or technology becomes an item of public concern or when the benefits of chemistry and chemical engineering are overneutralized by statements in the mass media which border on hysteria, the time has come for some sober and reflective action on the parts of all concerned—government, industry, and academe—as well as legal and socioeconomic students. Constructive action is in order to restore the sciences to their proper and absolutely essential place in our society's judgment.

This book is dedicated to the idea that sober persons, *acting in their own best interests,* will actively pursue the course of actions that will make Love Canals, Times Beaches, Stringfellow Acid Pits, Valley of the Drums, and other horror stories historical artifacts of the learning experiences based on the informed responsible actions of all concerned. We would turn around the figure reported in *Chemical & Engineering News* of the Cambridge Reports study, that 60% of the public believe there is no safe way and there never will be a safe way of disposing of chemical waste. In this book we outline such ways. The initiative of government, industry, and academe to move constructively and in concert is long overdue.

The widespread media interest in chemical and toxic (hazardous) chemicals, especially in the context of improper or inadequate disposal practices in the past, has seriously damaged the professions of chemistry and engineering, and has been a financial and public relations drain on all industry (since chemicals are used by *all* industry). This has spawned a "counterculture" movement to demote the science of chemistry from its prior image as a true and faithful servant of mankind, to a villain, lurking behind every drum and tank car with hidden dangers to the human race.

Phil J. Wingate, writing in *The Colorful DuPont Company* (Serendipity Press, Wilmington, Delaware, 1981, pages 7 and 8), has expressed it well:

"Chemicals are no longer thought of as magic bullets which would cure the ills of the world, but as poisons which are about to kill off all forms of life on earth. The term 'toxic chemical' has become almost one word, much like 'German dye' before WWI and 'damn Yankee' after the Civil War. Chemistry had ridden the roller coaster of public opinion from the heights of esteem to the depth of distrust.''

It is not the purpose of this author to criticize or to apologize. Certainly errors have been made in the past at all levels—industry, government, and academe. We appreciate the confusion in managements' and administrators' minds in accessing priorities, by the evolving political, legal, risk-benefit, and toxic tort threat proceedings, combined with the recent ''recession'' and the viewpoint which someone has labeled ''short-term profit mentality'' and the continuous progression of newer, less experienced personnel.

The subject of this book, hazardous and toxic wastes, is but one facet of the changing complex relationship between humans and science and technology. It has been noted by Samuel Flormar in *Blaming Technology: The Irrational Search for Scapegoats* (St. Martins Press, New York, 1982) that technology is not the reason for the changes, but the hope of the future. In the *Two Faces of Chemistry,* by Luciano Caglioti (MIT Press, Cambridge, Mass., 1983), the background for much of the misunderstanding regarding chemicals and their place in the world are reviewed. He concludes that we cannot go back, unless we are willing to accept a loss of life in unthinkable proportions. Going back to the beginning means reopening the door to epidemics and high infant mortality, giving up the use of herbicides and chemical fertilizers, and condemning to starvation hundreds of millions of people in addition to the many who are already starving. What he recommends is not to continue to advance indiscriminately nor stop or retreat on all fronts. We must face one problem at a time with honesty, intelligence, and humility. *Certainly the hazardous wastes and related problems are candidates for this approach.*

Although we are focusing on hazardous wastes, we are admitting our incomplete understanding of the whole process of manufacture, processing, transporting, and ultimate disposal of wastes. We *do* live in a chemical world, in which our lives are continually affected and nourished by chemicals in foods, drinks, transportation, housing, medicines, as well as in the water we drink and in the air we breathe. Most of these chemical substances and products are beneficial to our well-being. Unfortunately, as in all science, our knowledge base of hazards is never complete, what we know is poorly or incompletely disseminated, and occasionally we learn to our regret that some substances, both natural and manmade, must be given special attention and control. Wastes can be managed, reduced in volume, recycled, made less hazardous, and engineered to be assets instead of liabilities if we have the determination, as well as the legal and economic motivation to direct our attention to the task required.

I acknowledge with gratitude many who have been my mentors over the years, including Homer Carhart, Donald Katz, Glenn Seaborg, George H. Proper, Jr., and the late Floyd A. Van Atta.

George H. Proper, Jr. and Ms. Jean Miller were especially helpful in suggesting illustrations, and their assistance is much appreciated.

D. Jack Kilian, M.D. and Ms. Pamela Harris contributed Chapter 9 on medical surveillance, for which we are most grateful.

Certain parts of this book have been updated from the previous publication *Safety and Accident Prevention in Chemical Operations* (2nd ed., Wiley-Interscience,

New York, 1982), to which the reader is referred for a more complete and comprehensive treatment of chemical health, safety, fire, and radiation protection.

We are well aware that the billion dollar question remains:

> Oh, purge the land of toxic waste;
> Go sweep it, scrape it, pump it;
> But when the job at last is done,
> Where do you plan to dump it?
>
> Richard F. Barrett. *The Wall Street Journal,* August 18, 1983, p. 23.

However, perhaps this book will contribute some understanding, for, as Madame Marie Curie once said:

> Nothing in life is to be feared:
> It is only to be *understood*.

HOWARD H. FAWCETT

Wheaton, Maryland
April 1984

Contents

Hazardous and Toxic Materials

1

The Laboratory in the Beginning— The Edge of the Atom

Although chemical substances have been known, converted, and disposed of for thousands of years, starting with chemical oxidation, or fire, to produce heat, improve foods, and refine metals, only recently has serious attention been given to the potential benefit versus the risk involved. We shall begin by considering what we call "The Edge of the Atom"—considerations which we feel are essential to the constructive and proper utilization and disposal of the chemicals commonly used in the laboratory.

Chemicals are *neither hazardous nor safe* by themselves—it is only the improper utilization and disposal of chemicals or lack of controls that create problems. It must be stressed that chemicals, no matter what their state, their packaging, or trade names, are *centers of energy,* and *possess inherent properties* which should be considered if we are to benefit from their use, and avoid undesired side effects.

As Madame Curie pointed out, "Nothing in life is to be feared: It is only to be *understood.*" Control of chemicals cannot be intelligently undertaken until we truly understand. Chemical hazards sometimes are distorted or exaggerated out of reasonable or prudent prospective—which may be as unforgiving as indifference. To cite one recent example at a major university, 2 grams of thallium acetylide caused a 17-story building in which a chemistry laboratory was located to be evacuated for an entire day. The following is from the University of Massachusetts at Amherst *Environmental Health and Safety Bulletin,* and is included since it illustrates once again the need for careful preexperimental studies, and an awareness and emergency preparedness when conducting chemical research. Two organic chemists were attempting to bond terminal acetylenes to cyclopentadienyl systems for use in polymer research. They had patterned their experiment after analogous reactions in the literature, but which had used alkyl halides instead of acetylene. The reaction was to have gone as follows:

1.

2. This would next be added to iodotrimethylsilylacetylene, cooled down from $-78° \to 0°$, dissolved in tetrahydrofuran, and finally filtered.

3. After filtering, thallium ethoxide would be added.

4. Now any metal halide could be used to do the following, for example, $FeCl_3$.

What actually happened (or what is thought to have happened) is that reaction #2 never took place as the cyclopentadienyl thallium was too insoluble. So filtering merely removed the cyclopentadienyl thallium and left the iodotrimethylsilylacetylene. Now when the thallium ethoxide was added the following occurred:

$$CH_3-\overset{\overset{\displaystyle CH_3}{|}}{\underset{\underset{\displaystyle CH_3}{|}}{Si}}-C\equiv C-I + Tl^{\oplus}O^{\ominus}CH_2CH_3 \longrightarrow Tl^{\oplus}C^{\ominus}\equiv C-I$$

$$+ CH_3CH_2O-\overset{\overset{\displaystyle CH_3}{|}}{\underset{\underset{\displaystyle CH_3}{|}}{Si}}-CH_3$$

The thallium acetylide formed is a shock-sensitive, explosive compound. When the researcher touched a few milligrams of the substance in a hood with a spatula the substance exploded. As there were still a few grams of the product left, the researchers isolated the product in a hood, behind a protective shield, and called The Environmental Health and Safety Agency for aid.

Because the exact explosive and toxic potential of the thallium acetylide could not be positively assessed and because (from a security standpoint) partial evacuation of the General Research Center is very difficult, the entire building was evacuated until the Massachusetts Department of Public Safety Bomb Squad could come in and remove the compound.

Although in this instance no fault can be found with the students, the incident still raises questions about the safety of research work conducted on campus. It is the nature of research work to deal with unknown reactions.

To minimize possible hazards, effective controls based on relevant knowledge are necessary:

1. Find out as much as possible, before the experiment, about the particular procedure or reaction system (or a related one). Go to colleagues and research existing literature. Some good sources are:

   L. Bretherick, *Handbook of Reactive Chemical Hazards*, 2nd ed., Butterworth, London, 1979.

   H. H. Fawcett, "Safer Experimentation: Probing the Frontiers While Respecting the Unknown," *Safety and Accident Prevention in Chemical Operations*, H. H. Fawcett and W. S. Wood, Eds., Wiley–Interscience, New York, 1982, Chap. 22, pp. 421–488.

   Manual of Hazardous Chemical Reactions, 491-M, 5th ed., National Fire Protection Association, Boston, 1975.

   A. I. Vogel, *A Textbook of Macro and Semi-Micro Qualitative Analysis*, 4th ed., Longmans, London, England, 1976.

A. I. Vogel, *A Textbook of Practical Organic Chemistry,* 4th ed., Longmans, London, 1978.

A. I. Vogel, *A Textbook of Quantitative Inorganic Analysis,* 4th ed., Longmans, London, England, 1978.

2. Cautiously conduct a very small-scale preliminary experiment to assess the exothermic and physical properties of the reaction system and its products. When dealing with known explosive compounds (e.g., inorganic azides), quantities should be limited to less than 50 mg. For reactions with induction periods careful attention to the addition rate may not be sufficient, so a small-scale (25–50 ml) experiment should be run first.

3. Because the concentration of each reactant directly influences the velocity of the reaction and the rate of heat released, it is important not to use very concentrated solutions of reagents. Usually, a 10% concentration will suffice. But when using agents known to react vigorously, 2–5% may be more appropriate. For example, when concentrated NH_3 solution was used, instead of a diluted solution, to destroy dimethylsulfate, an explosive reaction occurred.

4. The rates of reactions increase exponentially with increases in temperature (roughly a 10°C increase will double the *rx* rate), so adequate cooling is necessary to control exothermic reactions. For example, explosive decomposition occurred when H_2SO_4 was added to 2-cyano-2-propanol with inadequate cooling—the exothermic dehydration of the alcohol produced methacrylonitrile, acid-catalyzed polymerization of which accelerated to an explosion.

5. Assess the hazard based on overall composition and detailed structure of the chemicals involved. In particular look at:

   a. *Overall Composition and Oxygen Balance of Mixtures and Compounds.* Oxygen balance is the difference between the oxygen content of a system and the oxygen required to fully oxidize the carbon, hydrogen, or other oxidizable elements present to CO_2, H_2O, and so on. In lab experiments, try to maintain the maximum negative oxygen balance (a negative balance means there is a deficiency of oxygen). To do this add oxidants very slowly and with appropriate cooling, mixing, and so on. Make sure that the desired reaction has in fact started, otherwise high concentrations of oxidant may accumulate before the onset of the reaction which, when finally started, may become uncontrollable. For example, oxidation of 2, 4, 6-trimethyltrioxane (paraldehyde) to glyoxal with nitric acid is subject to an induction period and may become violent if addition is too fast.

   In general, if the oxygen content of a compound approaches that necessary to oxidize the other elements present (with a few exceptions) to their lowest valency state, then the stability of that compound is suspect. Exceptions are nitrogen (which is usually liberated as a gas) and halogens (which go to halide if metal or hydrogen is present).

b. *Molecular Structure.* Instability and/or unusual reactivity in single
compounds is often associated with the following bond systems:

$C\equiv C$	Acetylenes, haloacetylenes, and metal acetylides
CN_2	Diazo compounds
$C-NO$	Nitroso compounds
$C-NO_2$	Nitro compounds
$C-(NO_2)_n$	gem-Polynitroalkyl compounds
$C-O-NO$	Alkyl or acyl nitrites
$C-O-NO_2$	Alkyl or acyl nitrates
$C=N-O$	Oximes
$C\equiv N \rightarrow O$	Metal fulminates
$N-NO$	N-Nitroso compounds
$N-NO_2$	N-Nitro compounds
$C-N=N-C$	Azo Compounds
$C-N=N-O$	Arenediazoates and bis-arenediazo oxides
$C-N=N-S$	Arenediazo sulfides, xanthates, and bis-arenediazo sulfides
$C-N=N-N-C$	Triazenes
N_3	Azides
$N=N-NH-N$	Tetrazoles
$C-N_2{}^+$	Diazonium salts
$N-C(N + H_2)-N$	Guanidinium oxosalts
N^+-OH	Hydroxylaminium salts
$N-Metal$	N—Metal derivatives (heavy metals)
$N-X$	N—Halogen compounds and difluoroamino compounds
$O-X$	Hypohalites
$O-X-O$	Halites and halogen oxides
$O-X-O_2$	Halates
$O-X-O_3$	Perhalates and halogen oxides
$C-Cl-O_3$	Perchloryl compounds
$N-Cl-O_3$	Perchlorylamide salts
$Xe-O_n$	Xenon–oxygen compounds
$O-O$	Peroxides
O_3	Ozone

Also included in this group are the aminemetal oxosalts, which are
compounds containing ammonia or an organic base coordinated to a

metal, with coordinated or ionic chlorate, nitrate, nitrite, nitro, per-chlorate, permanganate, or other oxidizing groups also present. For example, with heating, friction, or impact, dipyridinesilver perchlorate will decompose violently.

c. *Redox Compounds.* When the coordinated base in any of the above-mentioned compounds is a reductant (hydrazine and hydroxylamine) decomposition is extremely violent. For example, bis-hydrazinenickel (II) perchlorate has exploded while wet during preparation.

Other examples of highly energetic and potentially unstable redox compounds are those in which reductant and oxidant functions are close together in the same molecule, for example, hydroxylaminium nitrate.

d. *Pyrophoric Compounds.* These are compounds which are so reactive that contact with air and its moisture causes oxidation and/or hydrolysis at a high enough rate to cause ignition. When using these compounds an inert atmosphere and appropriate handling techniques are a must. Examples of these compounds are:

Finely divided metals—calcium and titanium.

Metal hydrides—potassium hydride and germane.

Partially or fully alkylated metal hydrides—diethylaluminium hydride and triethylbismuth.

Alkylmetal derivatives—ethoxydiethylaluminium and dimethylbismuth chloride.

Analogous derivatives of nonmetals—diborane, dimethylphosphine, triethylarsine, and dichloro(methyl)silane.

Carbonylmetals—pentacarbonyliron and octacarbonyldicobalt.

e. *Peroxidizable Compounds.* These compounds have been mentioned in previous newsletters. Basically, the common structural feature in organic peroxidizable compounds is the presence of a hydrogen atom which is susceptible to autoxidative conversion to the hydroperoxy group OOH. A few inorganic compounds and organometallic compounds can also form peroxides. Some typical structures susceptible to peroxidation are:

O—C—H	Ethers, cyclic ethers, and acetals
H₂C C—H H₂C	Isopropyl compounds and decahydronapthalenes
C=C—C—H	Allyl compounds
C=C—H	Vinyl compounds and dienes (i.e., monomers)
C—C—Ar H	Cumene, tetrahydronaphthalenes, and styrenes

Alkali metals, potassium, and sodium/potassium (NaK) alloys

Alkali metal,

Alkoxides and sodamide

Amides

It is important to test for, and if found, eliminate peroxides before using the compounds. Date peroxidizable compounds when first delivered to the lab so they can be disposed of when the recommended shelf life (3–12 months depending on chemical) is reached.

f. *Water-Reactive Compounds.*   Some classes of compounds which may react violently with H_2O (particularly a limited amount of H_2O) are:

Alkali and alkaline-earth metals (potassium and calcium)

Anhydrous metal halides (aluminium tribromide and germanium tetrachloride)

Anhydrous metal oxides (calcium oxide)

Nonmetal halides (boron tribromide and phosphorus pentachloride)

Nonmetal halide oxides (i.e., inorganic acid halides, phosphoryl chloride, sulfuryl chloride, and chlorosulfuric acid)

Nonmetal oxides (acid anhydrides and trioxide)

g. *Endothermic Compounds.*   While most chemical reactions are exothermic some are endothermic where heat is absorbed into the reaction product. These energy-rich products are thermodynamically unstable as no energy is necessary to decompose them and heat would be released. While some endothermic compounds with moderately positive values of standard heat of formation (e.g., benzene, toluene, and 1-octene) are not usually considered unstable, most endothermic compounds are unstable and possibly explosive.

Often the structure of the unstable endothermic compound involves multiple bonding, for example, acetylene, vinylacetylene, hydrogen cyanide, mercury (II) cyanide, dicyanogen, silver fulminate, cadmium azide, or chlorine dioxide. Hydrazine and dichlorine monoxide are examples of explosively unstable endothermic compounds without multiple bonding.

h. *Interactions of Oxidants With One or More Oxidizable Compounds and Interactions of Reductants With Materials Capable of Oxidation.*   In both cases the reaction may progress too fast to be controlled.

When it has been decided that a potential hazard exists in the proposed experiment, consult with your advisor. If your advisor agrees that you should proceed, take adequate precautions:

Use personal protective equipment (e.g., goggles, face shield, protective clothing, etc.).

Work in a fume hood behind barrier shields.

Have proper emergency equipment on hand [e.g., suitable fire extinguisher and emergency (SCBA) respirator if high-toxicity chemicals or cylinder gases are involved].

Notify the people who share the lab with you.

In another incident in a chemistry laboratory, as reported by Professor K. N. Raymond of the University of California at Berkeley to the Editor of *Chemical and Engineering News,* August 23, 1983, the synthesis of neodymium perchlorate from the metal oxide and perchloric acid was performed as described by Forsberg et al. [*Inorg. Chem.* **10,** 2656 (1971)]. The extractive purification of that material to make the anhydrous acetonitrile adduct, $Nd(ClO_4) \cdot 4CH_3CN$ [Eigenbrot and Raymond, *Inorg. Chem. 21,* 2867 (1971)], involves the Soxhlet extraction of neodymium perchlorate to give a solution of the anhydrous perchlorate acetonitrile. The removal of the solvent under vacuum at room temperature gives the tetraacetonitrile adduct, which although potentially explosive, has not been found to be shock sensitive. However, the drying of the acetonitrile extract at 80°C for approximately 24 hr generates a material whose composition corresponds approximately to the bis-acetonitrile adduct of neodymium perchlorate, $Nd(ClO_4)_3 \cdot 2CH_3CN$. This material is apparently very shock sensitive and approximately 2 g have detonated in our laboratories with tragic consequences. The research worker had momentarily removed his safety glasses, which resulted in the loss of one eye as well as two fingers.

In general, when noncoordinating agents are required, every attempt should be made to substitute anions such as $CF_3SO_3^-$, BF_4^-, and so on for perchlorate. If a perchlorate must be used, only small amounts of material should be prepared and they should be handled with adequate precaution.

The explosive power of perchlorates is well known. For example, ethyl perchlorate has five times the power of an equivalent amount of nitroglycerin. Perchloric acid hoods should be water-washed, and large installations require scrubbing of exhaust (see Fig. 1.1).

There is no lack of evidence for concern about chemical control. In Fig. 1.2 we have graphed the legal approach to controlling chemical hazards by showing the major chemical control measures which have been passed by Congress since the Food and Drug Administration (FDA) was established. Note especially the period 1970–1980, during which several major laws were passed. Each law was preceded by months of hearings and testimony. However, it should be noted that the legal approach is not always the effective means of controlling chemical hazards, since the implementation and enforcement of laws are not always the same as the law itself. In this context, note that federal agencies are required by an Executive Order to comply with the Occupational Health and Safety Act, P.L. 91-596.

Returning to fundamentals, we must consider the fundamentals as they relate to the control of chemicals. One important property of many common chemicals, especially of the solvents used in cleaning and related operations, is flammability. If this subject seems self-evident and elementary, the fact is that fires, explosions,

Figure 1.1. Perchloric acid fume scrubber system: Scrubber, neutralizing chamber, exhaust ducting and fume scrubber for perchloric acid hoods, and recirculating storage tank for alkaline absorber liquid. Photo by Center for Analytical Chemistry, National Bureau of Standards. Courtesy of Dr. J. Moody.

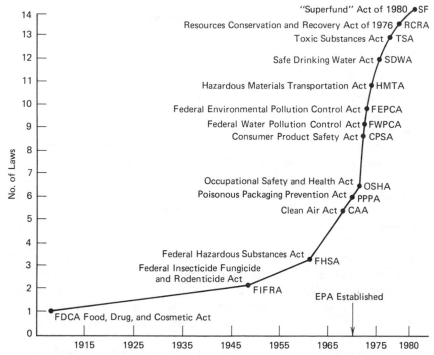

Figure 1.2. Federal laws pertaining to the control of chemicals in the United States. Note sharp rise in the numbers of laws since 1960.

and related deflagrations continue to be problems even in well-managed facilities. According to the National Fire Protection Association, more laboratories have suffered fire or explosion damage over the years than any other single type of chemically related accident. Historically, the classic description of fire was the triangle

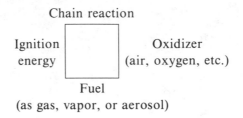

More recently we have come to recognize that a fourth parameter must be included, namely, the concept of the self-sustaining chain reaction, for a fire to continue. We now think of the fire as a square, as shown:

Given the proper combination of the four components, within critical limits, a fire or explosion is inevitable. What happens next depends on several factors, but even a small quantity, say a few milliliters of flammable solvents such as acetone or benzene, can start a serious fire. In turn, a small fire can lead to a conflagration. Proper handling and storage of flammable materials, as in metal cabinets, the use of carrying containers for materials to reduce breakage, and proper personal habits help reduce the risk of fire, as illustrated in Fig. 1.3.

Even substances not normally thought of as flammable will decompose under certain conditions. For example, in Fig. 1.4, note that trichloroethylene, usually considered nonflammable or noncombustible, has flammable limits of 14–40% by volume in air, assuming a high-energy ignition source. In the case of trichloroethylene, the pyrolysis or decomposition products include hydrogen chloride and phosgene, as well as carbon monoxide, making trichloroethylene degreaser fires or fume-offs both unpleasant and corrosive to metal. Many such incidents have in fact occurred.

Ignition can occur even if no obvious ignition source is present, if energy in the form of auto-ignition temperature (commonly referred to as AIT) is present. This is a measure of the energy needed to cause ignition or decomposition. Article 500 of the National Electrical Code, which specifies certain types of electrical equipment for hazardous areas or operations, is based on the understanding of this energy, and, by engineering, its prevention.

Another frequently overlooked laboratory hazard lurks where flammable materials are stored in a conventional domestic household refrigerator. Several ignition

Figure 1.3. Proper personal habits and respect for hazards are essential for control of chemicals in a human interface.

sources are present in, as well as outside, such boxes. Solvents evolve vapors which can be ignited by such sources. We first published this warning in 1950 in *Chemical and Engineering News,* but even today we occasionally hear of refrigerators exploding at reputable institutions. Explosion-proof refrigerators are now commercially available, as are instructions for removing the ignition sources from ordinary household boxes. Incidentally, the storage of chemicals and food or beverages in the same refrigerator is a highly questionable practice. Such boxes need frequent inspection, defrosting, and cleaning.

Although gases and vapors are most usually thought of in connection with fires and explosions, solids can burn and explode as well, such as grain dust explosions. In addition, several metals burn, especially when finely divided or when heated, such as magnesium, zirconium, uranium, hafnium, sodium, potassium, and Nak. Even iron, in the fine grade of steel wool, burns with surprising rapidity.

Another aspect of chemicals, like fire not completely appreciated, is toxicity. Toxicity is often equated to the word "poison," inferring that a substance may be toxic or nontoxic. That oversimplification leads to much misunderstanding. We prefer the definition: "Toxicity is a measure of the ability of an excess to damage life." You note the words *"excess"* and *"damage."* The fact is that the human body has a surprisingly high resistance and defense against insults from chemicals and other invasive substances. Built-in systems are quickly at work to neutralize,

LIMITS OF FLAMMABILITY
OF GASES AND VAPORS

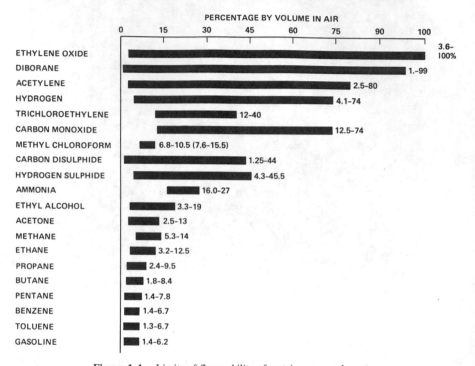

Figure 1.4.   Limits of flammability of certain gases and vapors.

metabolize, or eliminate materials, and, within limits, these systems are able to deal with a wide variety of exposures—not only with chemicals, as such, but with drugs including ethanol. The body has an intake defense system which can accommodate and effectively neutralize or reject significant doses of materials, whether they are chemicals or drugs. However, there is an upper limit as to how much can be tolerated how quickly, and for that measure we must refer to the language of the toxicologist. The toxicologist frequently reports the toxicity of chemicals in terms of LD_{50}, that is, the dose which will kill 50% of the exposed population (usually laboratory animals such as rats or mice). If inhalation exposures are reported, this is usually given as a LC or LC_{50}, the measure of the vapor or gas toxicity. It is by inhalation that most occupational exposures to chemicals occur, since we breathe a surprisingly large amount of air every minute, and the quick passage of the air into the lungs and then to the blood, ensures that prompt action will occur within the body system (see Chap. 6).

Chemicals are identified as having specific hazards to certain parts of the body. For example, inhalation of phosgene, the oxides of nitrogen, and ethylene chlorohydrin can produce pulmonary edema, a very serious effect which may occur several hours later, causing the body fluids to migrate or diffuse into the lungs, resulting in severe and often fatal respiratory problems.

Benzene, a commonly used solvent, is an example of both a flammable and a toxic liquid. In Fig. 1.5, we see the biologic metabolism of benzene. Note that the body metabolizes benzene into phenols and acids—highly toxic metabolates which are not easily rejected by the body. Substituted benzenes, such as isopropyl benzene or toluene or the xylenes, produce much less serious toxic effects when inhaled. All these, of course, pose serious hazards when inhaled in high concentrations. Benzene received much attention when the Supreme Court ruled that the Occupational Safety and Health Agency (OSHA) had not adequately shown that the risk/benefit of benzene justified drastic regulation of benzene as a chemical carcinogen. If there is doubt in your mind that benzene can be a problem, we suggest the current bestseller *Fever,* by Robin Cook, M.D., published in 1982 by Putnam. Although it is a novel, it contains very interesting observations on the problem of chemical hazard control and the medical and social implications.

An interesting aspect to the inhalation toxicity is the "carrier" potentiation. Although recently, according to the *Washington Post,* the U.S. Environmental Protection Agency (EPA) declared that formaldehyde is not a health hazard to humans, the data base does not agree. Especially interesting is the toxicity enhancement by concurrent inhalation of other substances. This concept of carrier enhancement and deposition may explain the recent observation that polonium 210 and lead 210 may concentrate in the lungs, emitting alphas in a localized area, following inhalation of cigarette smoke. Many manmade contaminants are present in the air we must breathe, and adding unnecessary pollution, such as cigarette smoke, to the mixture of contaminants, is clearly not in our personal interest. Indoor air pollution, especially from wood-burning stoves and fireplaces, have been studied sufficiently to suggest that the trade-off between economy and health is not always

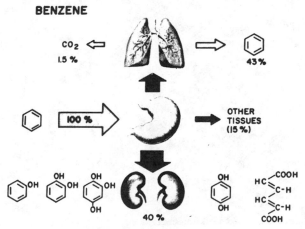

Figure 1.5. Benzene metabolism in the body. Note the phenolic and acidic structures which can produce serious biological effects, even in low concentrations.

a direct function. The U.S. EPA recently discontinued investigations into indoor pollution.

In a similar way, the use of nonprescription drugs, ranging from "pot" to more serious addictive drugs, clearly modifies the toxicity of chemicals to which we may be exposed, and causes supersensitivity from even minor occupational exposures.

Before finishing our discussion of toxicity, we should note constructive measures taken to prevent exposures to toxicity in the laboratory. One important approach is to position the work inside a properly designed, adequately maintained hood, such as illustrated in Fig. 1.6. Laboratory hoods are a complex subject beyond this book's restraints, but frequent monitoring is required by trained personnel so that hoods, through intelligent use and good housekeeping, are not used as storage areas or solvent-disposal media. (Note references in Bibliography at the end of this chapter.)

On a personal level, respirators are available for low concentrations of gases, dusts, aerosols, vapors, and other airborne exposures. The higher concentrations, as experienced by emergency personnel, fire departments, and others exposed to oxygen-deficient atmospheres, require the use of the self-contained breathing apparatus, preferably positive pressure, as shown on Figs. 1.7 and 1.8. The 1980 American National Standards Institute Z88.2 code on Respiratory Protection pre-

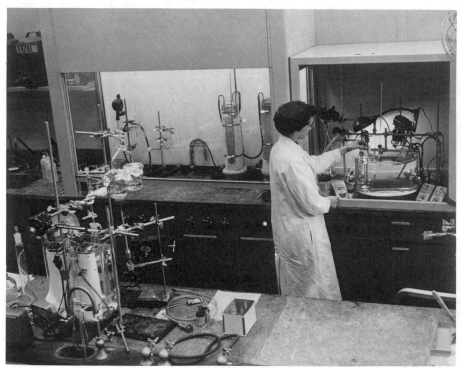

Figure 1.6. Properly engineered, monitored, and operated fume hoods of adequate capacity are essential to safe laboratory operations.

Figure 1.7.   Chemist uses proper protective equipment in spill clean-up. Courtesy of Smith Kline and French Laboratories.

sents the limitations as well as utilizations of respiratory protection and will be well worth the reading and understanding before using any respiratory protective device, in laboratory, plant, or waste disposal (see Chaps. 5 and 6).

If highly toxic work areas can be isolated, as shown in Fig. 1.9, and the worker fully protected by "pure breathing air" supplied by connections to the suit as well as to the breathing zone, substances which are otherwise too dangerous to handle can be handled in complete safety. This photo is from a "real-world" situation where a highly toxic dust is being handled. The protective suit has been carefully selected to ensure that it will resist penetration by the material in question. In choosing gloves and other protective equipment for chemical exposure, look carefully and critically at the data on resistance to the chemical in question, since no real test method, officially recognized, has existed over the years, although the American Society for Testing and Materials is nearing completion of a proposed test method for chemical penetration. It should be stressed, as depicted by Fig.

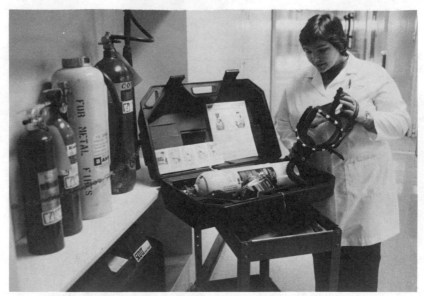

Figure 1.8.   Adequate training in location and use of emergency equipment is vital. Courtesy of Smith Kline and French Laboratories.

6.12, that no protective device is better than the sincere desire of the wearer to use it properly, and we would include eye, foot, head, as well as body, protection in this statement. Considerable literature is available on respiratory and other protection.

Several solvents, especially the chlorinated hydrocarbons, are known to produce serious effects on the liver and kidney organs, in addition to the narcotic or anesthetic effects from acute exposures. Carbon tetrachloride is a typical example of such a solvent. It is interesting to recall that not too many years ago, carbon tetrachloride was widely used as a cleaning solvent as well as in fire extinguishers, uses which are no longer recommended or condoned.

Although ingestion and inhalation are of great importance, the role of the intact skin in toxicity studies must be mentioned. In addition to the acids and alkalis, many of which are highly corrosive to the skin and cause rapid injury (such as the concentrated acids like sulfuric and nitric, and alkalis such as sodium hydroxide or potassium hydroxide), many substances may be adsorbed through the intact skin in sufficient concentrations to be toxic. The aromatic nitro and amino compounds are especially noteworthy in this connection, such as the nitrobenzenes and the aniline derivatives. To be practical, the use of proper protective gloves impervious to these substances; the frequent washing of the hands, arms, and face with warm water and soap several times a day; the wearing of long-sleeve clothing and long pants; and the use of appropriate eye or face protection is clearly recommended. The presence near the work area of a properly adjusted and easily reached safety shower, which will deliver a deluge of clean water at the rate prescribed by the American National Standards Institute on this subject, is a desired addition, as is the eye-wash fountain currently required by the OSHA regulations.

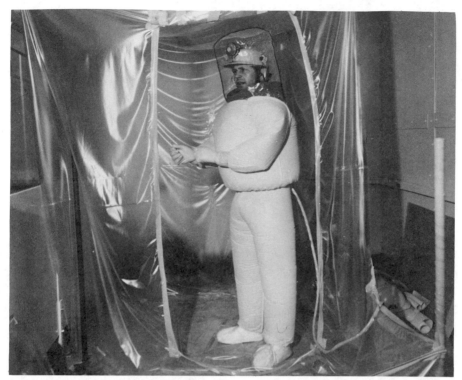

Figure 1.9. Worker protected in air-supplied suit and hood to prevent contact with highly toxic dust.

Before finishing our discussion on acids, an acid which deserves special caution is hydrofluoric. It poses a greater hazard than hydrogen chloride or hydrogen bromide solutions, since hydrofluoric acid (HF) is a protoplasmic poison even in dilute concentrations. Aqueous solutions as weak as 5% on the skin or in the eye can produce deeply penetrating burns, not always immediately obvious, which continue to penetrate very quickly even if washed off thoroughly on the surface. Burns of HF acid require *prompt* medical attention consisting of injections of calcium gluconate if proper treatment is to be encouraged, after thorough water washing. Medical personnel should be alerted that HF acid is in use, and should be encouraged to check their references.

It should also be noted that perchloric acid, as well as concentrated and fuming nitric acid, constitutes serious problems if allowed to mix with organic or easily oxidized materials, especially when heat is applied. The classic case involving perchloric acid-acetic anhydride mixtures used for the cleaning of aluminum in Los Angeles in 1947 attests to the strength of the explosive mixtures. Tests conducted by the author on mixtures of perchloric acid and ethyl alcohol, as well as nitric acid and ethyl alcohol, under conditions frequently encountered in metallurgical laboratories, conclude that extreme care must be taken to observe concentration and temperature considerations if fires or explosions are to be avoided (see Fig. 1.10). Even small amounts of exothermic materials must be treated with respect.

Figure 1.10.   Even small explosions of laboratory scale can produce significant hazard. Note one penetration of "shatter proof" hood window. Proper use of hood window prevented serious injury.

As noted above, chemical reactions are usually assumed to be beneficial and profitable, and they are when properly controlled, but to the prudent scientist, chemical boobytraps must be continually sought out and defined before they become real problems. As an example, two common household cleaners, namely, bleach (or sodium hypochlorite solution) and toilet bowl cleaner (sodium acid sulfate) when mixed evolve significant volumes of chlorine gas. The same evolution can occur with bleach and caustic (such as sodium hydroxide). Another example is the action of organic materials, such as sawdust, on nitric acid, to produce oxides of nitrogen, such as NO, NO_2, N_2O_4, and N_2O_5. A Boeing Pan American Cargo jet 707 crashed while attempting an emergency landing at Logan Airport in Boston, probably after the three crew members were overcome by the oxides of nitrogen from improperly packaged nitric acid on a flight from California to West Germany. Polymeric materials have been well studied by a National Research Council committee. Cabin materials in aircraft continue to be a problem when toxic gases are evolved in fires, illustrated by the 1973 crash landing of a Boeing 707 in which 122 persons died from the gases, and in 1980 when 301 died in an L-1011 cabin fire. Such fires can occur in waste-disposal areas, too. Many references are available, including the National Fire Protection Association 491-M manual on hazardous reactions, and the book, now in second edition, by Leslie Bretherick of BP Research Laboratory in England, which contains references to many hundreds of such reactions. A recent report on chemical compatibility, with particular reference to chemical-hazardous-waste disposal, is referenced in Chap. 7. It is important to become familiar with chemical compatibility and incompatibility, especially if we are to avoid future Love Canals.

Cryogenic fluids, such as the liquefied gases, liquid hydrogen, helium, nitrogen, oxygen, argon, methane, propane, and ammonia, deserve special mention since they possess unique hazards. The extremely cold nature of these fluids, clearly suggested by the temperature shown on the tanks, make skin and eye contact extremely painful. Another potential is the pressure build-up which can occur if the relief valves or outlet tubing on the container are plugged or restricted, such as might occur in transfer. It must be recalled that tremendous expansion occurs as the liquid becomes gas. Hydrogen, for example, expands 777 times, 1 liter of *liquid* hydrogen occupying 28.5 ft³ of *gas* at STP. Care to ensure that these containers are properly vented at all times will prevent overpressure. When filling containers on the floor, care must be taken to avoid splashed fluid from entering the shoe or sock of the person involved, and casual toeless shoes are clearly not recommended.

The chemical-disposal question is one that has only become of interest in recent times, since the Resources Conservation and Recovery Act of 1976 and the Superfund Act of 1980 began to be implemented (see Chap. 7). Both the U.S. EPA and the individual states have begun to control the transport and disposal even of laboratory quantities. The term "lab pack" has suddenly become popular, as a technique of packaging to permit wastes from laboratories to be safely and legally transported and disposed of. We must express our deep concern for the status of the waste disposal operations in this area (see Fig. 1.11). No longer can we think

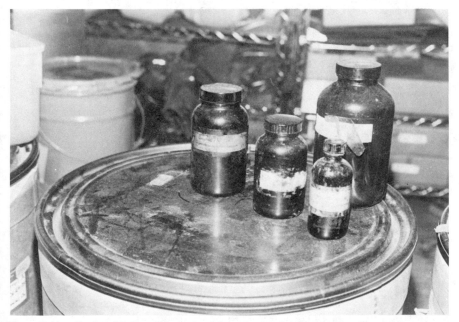

Figure 1.11. Careless labeling and storage of these suspected carcinogens in a biotest laboratory can invalidate test results as well as endanger lab personnel and complicate disposal procedures. Courtesy of H. H. Fawcett.

in terms of dumping chemicals down the sink drain, or in evaporating them up the hood. It is necessary to *identify* and *evaluate,* as well as *plainly mark* and *segregate* chemicals if they are to be safely, as well as legally, transported and disposed. Several identification systems exist.

A recent "real-world" observation by the author at an "approved," licensed chemical-waste-disposal site illustrates the problems. Two workers were attempting to sort out containers, totally without any identification, which originated in a major laboratory. One was a 2-lb peanut glass jar containing what we identified as metallic sodium. Since the first step used to reduce small quantities of chemicals is a jaw-crusher, such as used for crushing ice, over which a water spray was directed continually, it takes little chemical knowledge or imagination to visualize the explosion which was narrowly averted. Had this one jar "slipped through the system," newspaper headlines and television news would have recorded the situation. This is the "real world." Technical personnel have an especially important responsibility to ensure that their "wastes" are properly and adequately identified.

Laboratory-Waste Disposal

In view of the orientation of society to responsible control of hazardous materials, especially in disposal, the recent action of a committee of the National Research Council may be of interest to laboratory personnel and management.

Laboratories and factories both produce hazardous waste that must be disposed in an environmentally safe manner, but the similarity ends there. Industrial waste typically consists of large quantities of only a few compounds; laboratory waste generally includes small quantities of many different chemicals. Laboratories contribute only 0.1–1% of total hazardous waste in the United States, according to estimates by the EPA. Yet, laboratories are subject to the same regulations as industry.

A National Research Council committee investigating the problems laboratories encounter in disposing hazardous chemical wastes has proposed a variety of "safe, practical, and economically feasible ways" to dispose of laboratory chemicals. At the same time the committee concluded that the effects of some regulations "are out of proportion to the fraction of hazardous waste that laboratories generate and to the overall hazard it poses." The study included academic, government, and industrial laboratories.

The committee concluded that "hazardous chemicals can be handled safely and disposed of by environmentally acceptable methods," and it accepted the premise that the people who generate the waste have "moral and legal obligations" to see that it is handled and disposed of safely. But, added the committee, both handling and disposal "can be simplified without posing a hazard to health or a threat to the environment." One way to simplify would be for the EPA, DOT, and state and local governments—all responsible for some aspect of hazardous-waste disposal—to establish a "mutually consistent, interlocking regulatory approach."

One particularly onerous regulation for laboratories is a requirement for detailed reports on individual chemicals. To comply, a laboratory must keep extensive records of many chemicals, regardless of how small the amount handled—"an effort that seems unnecessary," the committee said. As an alternative, the committee suggested classifying compounds into seven categories reflecting their chemical characteristics: reactive, toxic, ignitable, corrosive (acid), corrosive (base), oxidizers, and miscellaneous. It proposed that compatible chemicals be permitted in one outer container—usually a 55-gal steel drum—and that records be kept according to classifications, not individual chemicals.

The committee identified three R's—reuse, recycle, recover—as the "first option" for handling wastes. When learned thoroughly and applied routinely, the three R's make clean-up safe, simpler, and cheaper.

Planning for disposal of leftover chemicals and products of chemical reactions should begin when the experiment is designed, the committee said. The goal should be to avoid difficult disposal problems whenever possible by reducing the amount of waste either through reclaiming chemicals or by using smaller quantities in the beginning of. The committee added that the experiment recycling also serves to reduce costs for new chemicals.

Chemicals still remaining after attempts at reusing, recycling, and recovering should be incinerated, if possible, or, as a last resort, burned in a properly engineered and approved landfill. Small quantities of certain hazardous chemicals sometimes can be converted in the laboratory to nonhazardous substances, easing disposal requirements (see pp. 203–205).

Prudent laboratory practices hinge on an institutional waste-management policy expressed in writing with specific procedures and assigned responsibilities. "It is essential," said the committee, "that laboratory management at all levels, or faculty in an academic institution, be openly and actively committed to support of sound waste-management policies and practices." "Equally important," the committee added, "is that all personnel who handle chemicals be thoroughly familiar with the lab's policy and with all regulations governing chemical handling and disposal."

The report covers all laboratory chemicals and biological waste that has been contaminated with chemicals, such as animal carcasses remaining from toxicological testing. Such waste requires disinfection or sterilization before disposal. The report does not address radioactive waste, controlled by other regulations.

Burning chemical waste is the cleanest way of handling virtually all organic compounds and some inorganic substances. However, incineration is not as easy as it may seem. Municipal incinerators usually are not equipped to handle hazardous substances, and currently available equipment designed for disposal of chemicals usually is too expensive for a laboratory to purchase and test for its own use. Commercial incinerators often refuse laboratory waste because it contains so many different chemicals. Furthermore, the incinerator ash still has to be disposed of in a landfill. The committee suggested that changes in regulations could provide greater incentive for the development of small, reasonably priced incinerators (such as pictured in Fig. 4.9, p. 82 of this book) for laboratory needs, and at the same time, could reduce the costs of operating an incinerator.

Most chemical wastes end up in landfills. Nontoxic substances may be accepted in a sanitary landfill operated by a municipality, but hazardous waste must go to a secure landfill, which the committee described as the "only practical option" at this time. This means the waste must be monitored for many years to protect ground water. Whether placed in a sanitary or secure landfill, costs can be lowered if the bulk of the waste package is reduced as much as possible before it leaves the laboratory.

Waste management is receiving considerably more attention than it ever had in the past. Landfills may be the cheapest course at the moment, but, as the public becomes more aware of the dangers of leaching chemicals and more Love Canals surface, less land will be made available and costs to use existing landfill will increase. "These factors may change the relative economics of recovery, recycling, and reuse." Meanwhile, laboratories are urged to continue to assess the three R's as a reasonable option.*

Only laboratory personnel can be expected to know what is involved or how to safely handle chemicals. Larger quantities are supposedly carefully analyzed and inventoried both by the state and the U.S. EPA regulations, and detailed records kept of the movement and ultimate disposal. Whether this actually occurs, we can only hope (see Fig. 1.12).

Much information is available for the technical personnel and scientists in the field of chemical health and safety, including fire. For example, toxicity and industrial hygiene is well covered by the recent update of *Patty,* in the form of five volumes of the third edition. Comprehensive data sheets for nearly 400 chemicals were issued recently by the NIOSH (National Institute for Occupational Safety and Health). More general treatments will be found, in the *Hazards in the Chemical Laboratory,* third edition edited by Leslie Bretherick for the Royal Society of Chemistry, and in *Prudent Practices for Hazardous Substances in the Laboratory* by a committee of the National Research Council. An excellent discussion on safe handling of chemical carcinogens in the laboratory is the ACS symposium volume edited by Douglas Walters. Academic laboratories, often populated by younger people, are covered by an increasingly large number of publications. There is no end to lists of hazardous chemicals. For example, the Consumer Products Safety Commission recently began a campaign to alert high-school instructors to the potential hazards of about 30 chemicals. Even the states, anticipating the New Federalism, are active. One state (California) recently issued a list of over 700 chemicals which it plans to, or in fact might already, regulate. The question of "safe exposures" for occupational exposures is addressed annually by a committee of the American Conference of Governmental Industrial Hygienists, and their recommendations are often cited into regulations for 8-hr-day exposures for a working lifetime for the average person. They hardly relate to the laboratory scene, except to give a rough index of relative hazard potential. The recent application of four-digit numbers as identification on bulk shipments such as you might encounter any day

---

*Condensed from *News Report,* Vol. XXXII, No. 7, September 1983, National Academy of Sciences, Washington, D.C., referring to the publication *Prudent Practices for Disposal of Chemicals from Laboratories,* ISBN No. 0-309-03390-X, (304 pp), 1983, available from National Academy Press, Washington, D.C. 20418.

Decision tree for classifying and disposing of lab wastes

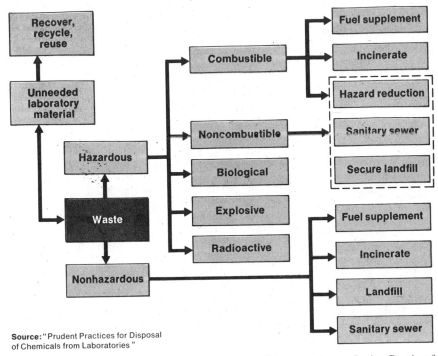

Source: "Prudent Practices for Disposal of Chemicals from Laboratories"

Figure 1.12. Decision tree for classifying and disposing of lab wastes. Source: Prudent Practices for Disposal of Chemicals from Laboratories.

on any highway or on the rails, the *Department of Transportation Emergency Response Manual,* now make it more difficult for anyone without the book at hand to identify the hazard potential of an overturned truck or rail car, and some of the suggestions in the DOT guide are highly questionable from a technical viewpoint.

In conclusion, we have reviewed some of the issues of chemical health and safety, pointing out that only a complete and comprehensive review of the literature, supplemented with new knowledge as it becomes available, can assure that we have the benefits of chemicals, while avoiding the sharp edges of chemical hazards. *From ordering reagents to the ultimate disposal, technical personnel have a legal and financial stake in understanding and follow-up of their chemicals. Identification, hazards, and location of chemicals should be known at all times, and the disposal should be made by competent personnel.*

BIBLIOGRAPHY

ANSI Standard for Emergency Eyewash and Shower Equipment, Z358.10198X, American National Standards Association, New York, 1981.

ANSI Standard for Precautionary Labeling of Hazardous Chemicals, Z129.1-1982, American National Standards Institute, New York, 1982.

Burgess, W. A., *Recognition of Health Hazards in Industry: A Review of Materials and Processes,* Wiley–Interscience, New York, 1981.

Caglioti, L. *The Two Faces of Chemistry,* MIT Press, Cambridge, Mass., 1983.

Denny, D., "Labeling Standard May Re-define Health and Safety Responsibilities," *Occup. Health & Safety,* 30–32 (January 1984).

Fawcett, H. H. "Safer Experimentation: Probing the Frontiers While Respecting the Unknown," and "Chemical Wastes: New Frontiers for the Chemist and Engineers—RCRA and SUPERFUND," in *Safety and Accident Prevention in Chemical Operations,* 2nd ed., H. H. Fawcett and W. S. Wood, Eds., Wiley–Interscience, New York, 1982, pp. 421–488 and 597–648.

Gerlovich, J. A., and Downs, G. E., *Better Science Through Safety,* Iowa State Univ. Press, Ames, 1981.

Green, A. E., *High Risk Safety Technology,* Wiley, New York, 1982, Chap. 2.1, Chemical, pp. 317–352.

Green, M. E., and Turk, A., *Safety in Working with Chemicals,* Macmillan, New York, 1978.

Katz, J., *Ozone and Chlorine Dioxide Technology for Disinfection,* Pollution Technology Review No. 67, Noyes Data Corp., Park Ridge, N.J., 1980.

Matheson Guide to Safe Handling of Compressed Gases, Matheson Division Searle Medical Products, USA, Secaucus, N.J., 1982.

National Safety Council, Chemical Section, *Six "Tip Sheets" on Safety in Chemical Industry,* 1982. (Available from the National Safety Council, 444 North Michigan Avenue, Chicago, IL 60611.)

Nowacki, P., *Health Hazards and Pollution Control in Synthetic Liquid Fuel Conversion,* Noyes Data Corp., Park Ridge, N.J., 1980.

Scottish Schools Science Equipment Research Center, *Hazardous Chemicals: A Manual for Schools and Colleges,* Oliver & Boyd, Edinburgh and Longmans, New York, 1980.

Wholen, J. P., "Using Emergency Wash Stations to Reduce Injuries," *Plant Eng.,* 79–82 (August 19, 1982).

Whythe, A. A., "Guide to Occupational Health Information Systems," *Occup. Health Safety,* 14–19, 36 (August 1982).

Wohileben, W., "Precautions Against Accidents in Chemical Facilities," *J. Hazard. Mater.* **5,** No. 1–2, 41–48 (1981).

Current Updates

Chemical Hazards, Health, and Safety, CA Selects (published every 2 weeks), Chemical Abstracts, 2540 Olentangy River Road, P. O. Box 3012, Columbus, OH 43210.

Chemical Information Abstract Service, BIT, International Labour Organization, 1211, Geneve, Switzerland.

Journal of Hazardous Materials (quarterly), since 1976 Elsvier, Amsterdam.

Laboratory Hazards Bulletin (monthly) Royal Society of Chemistry, The University, Nottingham NG7 2RD, England.

OSHA Reporter (weekly) published by Gershon W. Fishbein, 1097 National Press Bldg., Washington, D.C. 20004.

Publications of the *Bureau of National Affairs,* Washington, D.C.

Laboratory Safety

Fawcett, H. H., "Safer Experimentation," in *Safety and Accident Prevention in Chemical Operations,* 2nd ed., H. H. Fawcett and W. S. Wood, (Eds.), Wiley–Interscience, New York, 1982, Chap. 22, pp. 421–488.

Hazards in the Chemical Laboratory, 3rd ed., L. Bretherick, Ed., Royal Society of Chemistry, London, 1981.

National Academy of Sciences, *Prudent Practices for Handling Hazardous Chemicals in Laboratories,* National Academy Press, Washington, D.C., 1981.

"Readers Reveal the Dangerous Lives of R & D Scientists," *Ind. Res. Dev.,* 127–130, (July 1980).

Safety in Academic Chemical Laboratories, 3rd ed., American Chemical Society, Washington, D.C., 1980.

Management

"A Management System for Occupational Safety and Health Programs for Academic Research Laboratories," DHEW (NIOSH) Publication No. 79-121, January 1979.

"Occupational Safety and Health Program Guidelines for Colleges and Universities," DHEW (NIOSH) Publication No. 79-108, October 1978.

Chemical Health and Safety—General

Safety and Accident Prevention in Chemical Operations, 2nd ed., H. H. Fawcett and W. S. Wood, Eds., Wiley–Interscience, New York, 1982.

Benzene

Cook, R., *Fever,* Putnam, New York, 1982.

Toxicology and Industrial Hygiene

Patty's Industrial Hygiene and Toxicology, 3rd ed. (in 5 vols.), G. D. Clayton and F. E. Clayton (Eds.), Wiley–Interscience, New York, 1981–1982.

Encyclopedia of Occupational Health & Safety, 3rd Rev. Ed., International Labour Organization Geneva 22, Switzerland, 1983.

Laboratory-Waste Disposal

National Academy of Sciences, *Prudent Practices for Disposal of Chemicals from Laboratories,* National Academy Press, Washington, D.C., September 1983.

Laboratory Chemical Fume Hoods

"The Importance of Work Practices to Safe Hood Operations," 3/4-in. video tape, 20 min, prepared by Frank H. Fuller. [Available from Fairview Video, P.O. Box 128, Fairview Village, PA 19409. (Cost $25.00. Phone: (215)539-0769.]

See also Abstracts of Session on Laboratory Fume Hoods, 186th National Meeting, American Chemical Society, Washington, D.C., August 29, 1983. (Ten papers sponsored by the Division of Chemical Health and Safety of the ACS.)

Hospital- and Medical-Waste Disposal

Genetic Technology News. [Published by Technical Insights, Inc., 158 Linwood Plaza, P.O. Box 1304, Fort Lee, N.J. (monthly since February 1981).]

Recombinant DNA Technical Bulletin **6,** No. 2, (June 1983). (Published quarterly by National Institutes of Health, Bethesda, MD 20205.)

"Waste Disposal and Waste Disposal Systems," in *Engineering a Safe Hospital Environment,* D. L. Stoner, Ed., Wiley–Interscience, New York, 1982, Chap. 5, pp. 69–76.

Educational and Training Materials

CIS Abstracts (in English) of safety films, audiovisual materials, and training packages for use in occupational safety and health training of workers in chemical laboratories and processing operations involving chemicals. Information Search, Retrieval Service of the International Occupational Safety and Health Information Centre, Bureau of International du Travail, CH-1211, Geneve, 22, Switzerland, 1982.

Freeman, N. T. and Whitehead, J., *Introduction to Safety in the Chemical Laboratory*, Academic Press, London, 1982.

National Research Council, *Indoor Pollutants,* National Academy Press, Washington, D.C., 1981. Discusses indoor combustion, formaldehyde, tobacco smoke, radon, asbestos, microorganisms, allergens, temperature, and humidity.

Aircraft Safety

"Technology," *Wall Street Journal,* February 19, 1982, p. 31.

16-mm Motion Pictures

Chemical Boobytraps, 1959, (10 min.) (Directed by H. H. Fawcett, General Electric Research Laboratory, Schenectady, N.Y.).

Hazardous Waste Options, 1981, (28 min). (Made for the EPA by Stuart Finley of Falls Church, Virginia.)

Hazardous Reactions

Bretherick, L., *Handbook of Reactive Chemical Hazards,* 2nd ed., Butterworths, London, 1979.

Manual of Hazardous Chemical Reactions, 491-M, National Fire Protection Association. (Available from the NFPA, Batterymarch Square, Quincy, MA 02169.)

Indoor Air Pollution

Survey of Indoor Air Quality Health Criteria and Standards. (Prepared for the EPA by Geomet, 1801 Research Boulevard, Rockville, MD 20850.)

Hicks, J. B., "Tight Building Syndrome: When Work Makes You Sick," *Occup. Health & Safety,* 51–57 (January 1984) (6 refs).

Health and Medical Aspects

Proctor, N. H., and Hughes, J. P., *Chemical Hazards of the Workplace,* Lippincott, Philadelphia, 1978.

"Mobile Health Testing," *Occup. Health & Safety,* 75 (January 1984).

Classification of Hazardous Substances

"A Summary of Hazardous Substance Classification Systems," EPA/530/SW-171, December 1975. (Available from the U.S. Environmental Protection Agency, Cincinnati, OH 45268.)

Professional Affiliation

Division of Chemical Health and Safety, American Chemical Society, Douglas Walters, c/o National Toxicology Program, National Institute of Environmental Health Science P.O. Box 12233, Research Triangle Park, NC 27709.

Division of Safety and Health, American Institute of Chemical Engineers, 345 E. 47th St., N.Y., N.Y. 10017.

Journals

Journal of Hazardous Materials. (Published quarterly by Elsvier, Amsterdam.)

Journal of the American Industrial Hygiene Association. (Address: 475 Wolf Ledges Parkway, Akron, OH 44311.)

Occupational Health & Safety, Medical Publications, Inc., 5002 Lakeland Circle, Waco, TX 76710 (monthly).

Publications of the Chemical Section and R & D Section, National Safety Council. (Address: 444 North Michigan Avenue, Chicago, IL 60611.)

Publications of the National Fire Protection Association. (Address: Batterymarch Square, Quincy, MA 02269.)

2
Toxicity—Part 1

The wide interest in toxicity, especially as the NIOSH, OSHA, FDA, and EPA have approached the subject, might suggest that some new aspect of chemicals had recently been discovered. In fact, the ability of chemical substances to affect human and animal life was known to ancient man. Alchemists of the Dark Ages, who were the progenitors of modern chemical science, were both seeking the Philosopher's Stone (with which to bestow eternal life on mortals), as well as a method for converting base metals, such as lead, zinc, and mercury, into gold, the uncorruptible metal. Several important drugs were first isolated from roots, flowers, or other parts of plants used for centuries by Indian tribes. For example, curare and digitalis, both powerful drugs in common use today, were first isolated from natural sources. An understanding of toxicity is an important step towards the safe and proper utilization of drugs and chemicals.

The emphasis that has been placed on toxicity frequently overlooks the fact that even today our knowledge base of toxicity is limited and fragmented. Society, through consumer groups and various governmental agencies, especially the FDA, CPSA, NIOSH, OSHA, Department of Agriculture, DHHS (which includes the National Cancer Institute, NCI), the National Clearinghouse of Poison Control Centers, and the EPA, under the Toxic Substances Control Act of 1976 (P.L. 94-1302) as well as local and state interests, has fragmented the research and control activity, and we face a long period before the emotional and economic aspects of toxic substances control are resolved. It would be unfair to neglect to mention the outstanding toxicity and safety evaluation work conducted and published in the open literature over the years by several major chemical companies, including Union Carbide, Dow, DuPont, and Eastman Kodak, long before public concern or legal requirements were at their present level. The Chemical Industry Institute of Toxicology, which has sponsorship from major companies, was established in 1975 and has underway an ambitious program of toxicological research and testing of commercial chemicals.

Toxicity is not a strange unknown. Rather, it is a measure of the action of a given amount of a chemical or other substance, on a designated living species or system (plant or animal), to quantify the level or dose at which the system is damaged beyond prompt recovery when that dose enters and is absorbed by the living system. Even an element which is absolutely essential to human life, namely, oxygen, is toxic under conditions of prolonged inhalation at elevated pressure, as in scuba diving below 32 ft (9.76 m). Several elements essential to life in low concentrations, such as zinc, copper, magnesium, selenium, and arsenic, are toxic or lethal at higher concentrations. Hence, to simply label a substance "toxic" or "nontoxic" without specifying the details and limits of the data, is misleading. Note that until the dose has actually entered the body and reached critical organs by one or more routes, such as inhalation, ingestion, and skin absorption, toxicity will not be manifest (see Chap. 3).

Unlike the many values pertaining to a chemical tabulated in handbooks and critical tables, toxicity is not a specific property or physical constant derived by simple physical measurement in the abstract. Because it depends on measurements relative to living systems, it is never absolute, since individual members of all living species and systems, especially humans and animals, vary among themselves and with other species in their resistance or susceptibility to insults from toxic materials. Defense mechanisms, by which humans and other hosts adapt or resist a toxic agent, include adaption, response to stress, and complex biochemical mechanisms which act to minimize or prevent toxicity. When these homothesis actions are overburdened, toxic effects are always observed.

We recall one such response "escape" measure we observed in connection with exposures to oxides of nitrogen, a not infrequent emergency and well recognized by experienced operators for serious or fatal consequences. When a nitrator operator suddenly saw a cloud of dark brown or orange-colored fumes in his immediate area, he would instinctly bring his operation under control, and then escape—first, by holding his breath, then by "shallow breathing," that is, by restricting his breathing to a few hundred milliliters of tidal volume. He can thereby keep his respiratory system operating, avoiding hypoxia (oxygen deficiency), but not inhale significantly the fumes into the lower respiratory tract, where damage to the alveoli could occur and produce serious or fatal pulmonary edema, often several hours later. The urge to "panic" and "run" was suppressed. Shallow breathing permitted survival in an otherwise toxic environment until escape, or allowing sufficient time to a respiratory protective device. The recently reported "low-temperature" process for nitrocompounds, operated from a remote control room, has hopefully made exposures to oxides of nitrogen less frequent, but any person involved in operations the environ of which suddenly contains any highly toxic gases may profit from the survival technique of shallow breathing. (See L. Albright and C. Hanson, *Industrial and Laboratory Nitrations,* ACS Symposium Series 22, American Chemical Society, Washington, D.C. 1976.)

Humans and animal systems have several other mechanisms for dealing with insults from toxic substances, often called detoxification. At some point on the dose-effect (or dose-response) curve, however, acute or chronic injury can occur—which may or may not affect every member of the subject group, or be immediately obvious. The "injury" may be manifest in one or several ways, such as modified or irregular behavior (intoxication), central-nervous-system disorientation, impairment of body functions, allergic reactions, skin and eye irritation or damage, respiratory and cardiac collapse, pulmonary edema, or by less obvious delayed reactions which may require years before producing carcinogenic, mutagenic, or teratogenic effects in essential organs. Until the toxic dose actually enters the body through one or more routes of entry, such as inhalation, ingestion, or cutaneous (skin) absorption, it is probably harmless. The possible exception to the previous statement is the possibility that a one-hit injury may occur; that is, one molecule of the substance contacts and damages the essential DNA and RNA of a critical organ, and may induce carcinogenic activity. This theory of no threshold for exposure to chemical carcinogens is widely debated, and impressive arguments may be cited both for and against this theory. It is possible that our understanding of this process is so incomplete that a threshold exists, but has not yet been recognized and accepted. However, for most chemicals, the consensus is that a threshold value exists, based on protection for the majority of the workers who will be exposed, but the degree of protection it affords any one individual in a group is not clear at this writing.

Since the original presentation of a list of threshold-limit values by Professor Warren Cook in 1945, many changes have occurred annually, as the list was revised by the American Conference of Governmental Industrial Hygienists (ACGIH). The 1969 threshold-limit values published by the ACGIH became the OSHA standards for most substances [see *Fed. Reg.* **36,** No. 105 (May 29, 1971)]. Since then, the annual revisions have continued by the ACGIH and have been expanded to cover over 600 substances, as well as to physical agents including light, noise, lasers, microwaves, ionizing radiation, and heat stress. As noted in the preface to the ACGIH compilation for airborne contaminants in the workplace, to which the reader should refer for the most current values, three categories of threshold-limit values (TLVs) are specified.

1. *Threshold-Limit Value—Time-Weighted Average (TLV–TWA).* This is time-weighted average concentration for a normal 8-hr workday or 40-hr workweek to which nearly all workers may be repeatedly exposed, day after day, without ill effect. To cite a few typical values as examples from the 1983–1984 listing:

Acetone	750 ppm or 1780 mg/m^3	(TLV–TWA)
Acrolein	0.1 ppm or 0.25 mg/m^3	(TLV–TWA)
Ammonia	25 ppm or 18 mg/m^3	(TLV–TWA)
Carbon monoxide	50 ppm or 55 mg/m^3	(TLV–TWA)
Triethylamine	10 ppm or 40 mg/m^3	(TLV–TWA)
Zinc oxide fume	5 mg/m^3	(TLV–TWA)

2. *Threshold-Limit Value—Short-Term Exposure Limit (TLV–STEL).* This is the maximal concentration to which workers can be exposed for a period of up to 15 min continuously without suffering from (1) intolerable irritation, (2) chronic

or irreversible tissue change, or (3) narcosis of sufficient degree to increase accident proneness, provided that no more than four excursions per day are permitted, with at least 60 min between exposure periods, and providing the daily TLV–TWA also is not exceeded. The STEL should be considered a maximal allowable concentration, or ceiling, not to be exceeded at any time during the 15 min exposure period. Examples of the TLV–STEL for the same substances as above in the 1983–1984 listing are:

Acetone	1000 ppm or 2375 mg/m^3	(TLV–STEL)
Acrolein	0.3 ppm or 0.8 mg/m^3	(TLV–STEL)
Ammonia	35 ppm or 27 mg/m^3	(TLV–STEL)
Carbon monoxide	400 ppm or 440 mg/m^3	(TLV–STEL)
Triethylamine	15 ppm or 60 mg/m^3	(TLV–STEL)
Zinc oxide fume	10 mg/m^3	(TLV–STEL)

3. *Threshold-Limit Value—Ceiling (TLV–C).* This is concentration that should not be exceeded even instantaneously. Examples of the TLV–C from the 1983–1984 values are:

Boron trifluoride	Ceiling value of 1 ppm or 3 mg/m^3
Hydrogen chloride	Ceiling value of 5 ppm or 7 mg/m^3
Manganese dust and compounds, as Mn	Ceiling value of 5 mg/m^3

For some substances, such as irritant gases, only one category, the TLV–C, may be relevant; for others, the TLV–TWA and TLV–STEL may be appropriate. In the compilation of the ACGIH, substances which are known to be toxic through the intact skin are so designated by the addition of the word "skin" after the chemical name. As an example, aniline, one of several substances which can be absorbed through the intact skin as well as by inhalation is listed:

Aniline—skin   TWA: 2 ppm or 10 mg/m^3 STEL: 5 ppm or 20 mg/m^3

From the discussion of threshold-limit values, it is obvious that the subject of exposure limits is complex. Therefore, the warning quoted from the preface to the ACGIH compilation is clearly in order:

These limits are intended for use in the practice of industrial hygiene and should be interpreted and applied only by a person trained in this discipline. They are not intended for use, or for modification for use, (1) as a relative index of hazard or toxicity, (2) in the evaluation or control of community air pollution nuisances, (3) in estimating the toxic potential of continuous, uninterrupted exposures or other extended work periods, (4) as proof or disproof of an existing disease or physical condition, or (5) for adoption by countries whose working conditions differ from those in the United States and where substances and processes differ.

In experiments on humans using radioactive tracers in pesticides, and using the forearm as the frame reference, it has been shown that the palm, of which the thick stratum corneum is allegedly almost impenetrable, allowed approximately the same penetration as the forearm. The abdomen and dorsum of the hand had twice the penetration of the forearm. The follicle-rich sites, including the scalp, angle of the jaw, postauricular area, and forehead had fourfold greater penetration. The intertrigenous axilla had a fourfold to sevenfold increase; the scrotum allowed almost total absorption. The overall impression is that all anatomic sites studied show significant potential for penetration of pesticides, and hence systemic intoxication. [See H. I. Maibach et al., "Regional Variation in Percutaneous Penetration in Man," *Arch. Env Health* **23**, 208–211 (September 1971).]

In addition, the degree of toxicity is influenced by:

Dose.   Generally the larger the dose (or higher the concentration), the more rapid the action

Rate of Absorption.   The faster this rate, the quicker the action. With oral administration, the intoxicating and the lethal doses may be considerably influenced by the condition of the gastrointestinal tract, especially by the amount of food and other matter in the stomach and intestine. A vehicle, such as oil, also affects absorption of a skin exposure.

Route of Administration.   For the most part, toxicity is greatest by the route that carries the toxic substance to the bloodstream most rapidly. In *decreasing* order of speed of action, routes for most drugs and other substances are (see Chap. 3):

  Intravenous
  Inhalation
  Intraperitoneal
  Intramuscular
  Subcutaneous
  Oral
  Cutaneous (skin)

Food in the alimentary canal may delay or decrease toxic action. Digestive enzymes may destroy or alter the compounds with resultant changes in the toxicity. Certain compounds are virtually harmless if taken orally, but lethal when introduced parenterally; in many other cases, the converse is true. The toxicity of the material may also vary considerably with the form in which it is administered, for example, solid, in suspension or in solution. In solution, the toxicity may be influenced by the solvent and the concentration. Synergism (or combined effect) may be a very real factor in the action, and probably occurs far more frequently than recognized. In the area of "mixed exposures," knowledge is very uncertain. As an example, exposures of inhalation of hydrogen sulfide, or to trichloroethylene, followed by ingestion of alcohol, greatly increases the effects of both.

Therèfore the toxic dose is influenced by:

Site of Introduction. With subcutaneous injections, toxicity may be affected by the density of the subcutaneous tissue. With intravenous administration whether the injection is made into the femoral or jugular vein may be of importance, but in any case the rate of injection, or the amount of material injected per unit time, will considerably influence the value of the toxic dose.

Other Influences. Disease, environmental temperature, habits and tolerance (such as smoking, alcohol intake, other drugs, low-level exposure to carbon monoxide, and other substances), idiosyncrasy or allergy, diet, and season of the year all may influence the toxicity of a substance. The toxicity of chemicals also vary with the species of animals used, and sometimes with different strains of the same species. Within the same strain significant differences between litter mates have been observed. This variation is not unique to laboratory animals, but can be observed in humans as well. As Kubias has pointed out, there is no average person.

Since in the real world people have a wide variety of exposures, from the moment of birth into the air of this unclean world, and the defense processes which each person possesses may change from time to time, it is seldom possible to attribute the toxic effects of any substance specifically without consideration of the other forces from other exposures which are operating over the years. In discussing the defense processes which are often the determining factor in human reactions to a chemical, G. R. C. Atherley, in *Occupational Health and Safety Concepts,* published by Applied Science Publishers, Ltd., London, 1978, notes respiratory filtration of aerosols and mists (which we consider in Chap. 6), cellular defense, inflammatory response, immune response (which alters sensitivity), homeostasis (in which the internal milieu of the body is maintained in a stabilized condition), stress resistance (linked to the hormone cortisol), thermoregulation of body temperature, and metabolic transformation whereby a toxic substance is rendered less toxic and can be excreted by the kidneys.

Because of these defenses, which are most effective in healthy adults, the epidemiologists, when studying a large number of persons of a given population, such as employment on similar occupational tasks for a comparable period of time, often note that they are usually the population with the highest resistance, due partly to preselection, to excellent medical attention, to adequate and balanced diet, and other factors which are in favor of good general health.

Although a direct relationship between an occupation (chimney sweeping) and cancer was established by Potts in 1775, the subject of potential carcinogenic activity of common chemicals did not receive widespread attention until fairly recent times. Several studies have suggested that between 50 and 80% of human cancer has its origin in occupational or ambient (real-world) environments. Since 1970, when it became a stated policy of the government to develop the information necessary to eliminate cancer from its position as one of the leading causes of adult death in the United States, various approaches have been made to the problem. The National

Cancer Institute (NCI), which is part of the National Institutes of Health, has been studying cancer both by epidemiology studies of human populations exposed, and from studies in animals, beginning with rats and mice, and going up the scale to nonhuman primates. During the past several years and starting in 1972, several hundred chemicals and chemically related substances have been tested, either by the NCI itself, or through grants and research contracts. Unfortunately, most rodent studies require 2 years for the experiment to be completed, and usually another year before the data is fully tabulated and reported in acceptable form. Information on the availability of information from testing may be obtained from the Office of Information, National Cancer Institute, Bethesda, MD 20205, Attention: Information Specialist, or call (301) 496-6095/496-5583.

Reviews of the older literature have shown that perhaps 2000 chemicals are suspected carcinogens, based on data currently available.

Recognizing that the dissemination of information to workers is an important factor in educating persons to avoid contact with toxic substances, or, if they prefer, to seek alternate employment, the Occupational Safety and Health Administration requested the Assembly of Life Sciences of the National Research Council to recommend how to inform workers. The committee concluded that when controlled experiments demonstrate that an agent produces cancer in animals, that agent should be regarded as possibly carcinogenic in humans. It recommended that a single national source determine if an agent is of sufficient hazard to warrant informing workers. This source, such as the Department of Health and Human Services, should be one that is, to the extent possible, credible to both management and labor and one that is not involved in the regulatory process. The committee noted that apparently effective employee information programs conducted by large corporations and associated labor unions do exist. However, it expressed concern with the large number of workers who might be exposed to carcinogens in small plants. These employees would not have access to the health staffs and other benefits of larger plants, nor to the resources of a union. For that matter, the very existence of their workplaces would probably not be known to the government agencies that sought to inform them. At least 75% of the nonagricultural work force is not affiliated with labor unions. Plant size varies considerably. For example, in 1973, the construction industry employed almost 600,000 people who could have been exposed to asbestos in their working environment, spread out over about 70,500 establishments. Only 2% of these were employers of more than 50 workers each. Fifty-one percent employed fewer than four workers each. [The EPA is moving away from the notion that everything is assumed bad for human health unless specifically proven to be otherwise. (See "Administration Takes Steps to Ease Control on Cancer-Causing Agents," *Wall Street Journal,* December 24, 1982, p. 7.) In August 1982, the Manville Corporation, under Chapter 11 of the Bankruptcy Act, filed to shield its assets from 14,000 claims alleged due to excessive asbestos inhalation.) (See also B. Richards, "UNR Case May Set Legal History with Effort to Limit Asbestos Suits," *Wall Street Journal,* March 16, 1983, p. 33.) On August 26, 1983, the OSHA announced it was considering an emergency rule for a limit of 500,000 asbestos fibers from 2 million fibers per cubic meter of air (see *Wall Street*

Journal, August 26, 1983, p. 1 and 2.) The OSHA estimates that 3 out of 1000 workers exposed at the higher level for 1 year will subsequently die from that exposure.]

The use of bacteria as an indicator of both carcinogenic and mutagenic potential has received much attention in the past few years. The Ames test, which was developed by Professor Bruce Ames at the University of California, uses the Salmonella microsome, and certain other recognized test strains, as the test subject. The test can be run and reported in a few days at nominal cost. Professor Ames estimates that the cost for a compound would be in the order of $100–$300, and one person could do several compounds a day. The acceptance of the Ames approach has set the stage for development of new and ingenious techniques that, used together, may provide accurate information on the possible hazards of chemicals. Six short-term tests for detecting carcinogenicity have been evaluated using 120 compounds, half of which were carcinogens and the remainder noncarcinogens. The results obtained indicate that the Ames test and a "cell transformation" assay are both sufficiently sensitive to carcinogenicity, or the lack of it, in the compounds studied to enable them to be employed for detecting potential carcinogens, if used under carefully controlled conditions.

At a conference held at Cold Spring Habor, Long Island, New York in September 1976, a large assembly of scientists from many disciplines considered the problem of carcinogenesis, and the best approach to resolving acknowledged unknowns or uncertain aspects of our present knowledge. They reaffirmed what was mentioned previously, namely, that humans *are* animals in this context, and that test data that suggest or show that a certain chemical produces neoplasms in lower animals can be used as predictive for humans. Citing specific cases where well-documented animal data were available that proved applicable to humans, Dr. Rall of the National Institute of Environmental Health Sciences mentioned that in 1945 the data showed that diethylstilbestrol (DES) produced cancer in mice. Since the medical profession prescribed widespread use of this substance as a female hormone supplement in early pregnancy, it has been shown to have produced vaginal cancer in the daughters of some women. Likewise, 4-aminobiphenyl was studied in rats and dogs from 1952 to 1954, and shown to produce bladder cancer. Production of this compound in the United Kingdom never was attempted. In the United States, starting in 1935, the compound was produced for 20 years. Today it is one of the "OSHA carcinogens."

As Dr. Lorenzo Tomatis of the International Agency for Research on Cancer in Lyon, France has pointed out, the prevailing preoccupation in the research community is "not to extrapolate unduly for experimental data to man," yet very little emphasis, if any, is put on the mistakes made when results obtained in animals were not taken as indicative of a possible danger to man.

A concensus reported from the Cold Spring Harbor conference included:

1. The chain of events which converts normal healthy cells into irreversible carcinogenic neoplasms is still incompletely understood.

2. Really effective new cancer treatments are not on the horizon.

3. Prevention—the only long-range "cure"—is still not understood or applied to any real degree. (Cigarette smoking and auto exhausts containing known carcinogens, may be cited as knowns about which little is being done, even today.)

We can offer the personal opinion that, as public and social pressures increase, more aggressive action will be forthcoming. The Toxic Substances Control Act (TSCA) (signed by President Ford on October 1976 and effective as of January 2, 1977) is an example of such concern reduced to law, which hopefully will be administered wisely and intelligently, in the national interest. [The health-care costs from lung cancer associated with cigarette smoking has been estimated at $17 billion annually for the United States alone. Public awareness of the high cost of smoking is having real impacts. (See, e.g., N. Thimmesch, "No Smoking: No Ifs and Butts," *Saturday Evening Post,* July/August 1983, pp. 30–35, and "Part II, No Smoking Signals from the Work Place," *Saturday Evening Post,* Volume 255, Number 6, September 1983, pp. 36–40.) Another example of a growing health problem is alcoholism. Estimates of the number of alcoholics in the United States range as high as 15 million. Approximately 67% of the adult population in the United States uses alcohol on occasion and 12% are considered "heavy drinkers." Hepatic (liver) disorders are the major medical finding, since in the body alcohol is oxidized to acetaldehyde and hydrogen equivalents. Little appreciated is the fact that each gram of ethanol provides 7.1 cal; 12 oz of an 86-proof beverage contains 1200 cal, half of the recommended daily dietary allowance. (See C. S. Lieber, "Hepatic, Metabolic, and Nutritional Complications of Alcoholism," *Res. Staff Phys.,* 79–96 (August 1983) (15 refs.).]

2.1. THE TOXIC SUBSTANCES CONTROL ACT OF 1976 (P.L. 94-469)

After 5 years of congressional hearings, study, and debate, the Toxic Substances Control Act of 1976 is now law. Before outlining the law's major sections, the Findings of Congress and Policy of the United States, as presented in Section 2 of the law are pertinent:

(a) *Findings: The Congress finds that—*

(1) human beings and the environment are being exposed each year to a large number of chemical substances and mixtures;

(2) among the many chemical substances and mixtures which are constantly being developed and produced, there are some whose manufacture, processing, distribution in commerce, use, or disposal may present an unreasonable risk of injury to health or the environment; and

(3) the effective regulation of interstate commerce in such chemical substances and mixtures also necessitates the regulation of intrastate commerce in such chemical substances and mixtures.

(b) *Policy: It is the policy of the United States that—*

(1)   adequate data should be developed with respect to the effect of chemical substances and mixtures on health and the environment and that the development of such data should be the responsibility of those who manufacture and those who process such chemical substances and mixtures;

(2)   adequate authority should exist to regulate chemical substances and mixtures which present an unreasonable risk of injury to health or the environment, and to take action with respect to chemical substances and mixtures which are imminent hazards; and

(3)   authority over chemical substances and mixtures should be exercised in such a manner as not to impede unduly or create unnecessary economic barriers to technological innovation while fulfilling the primary purpose of this Act to assure that such innovation and commerce in such chemical substances and mixtures do not present an unreasonable risk of injury to health or the environment.

The law requires the EPA to order animal testing of any chemical or mixture for which data and experience are insufficient, and which are relevant to a determination that the manufacture, distribution in commerce, processing, use, or disposal of each substance or mixture, or any combination of such activities does or does not present an unreasonable risk of injury to health or the environment. The standards to be prescribed for the testing of health and environmental effects will include carcin-

Figure 2.1.   Scientist checks on inhalation studies in animals. Courtesy of Dow Chemical Co.

ogenesis, mutagenesis, teratogenesis, behavioral disorders, cumulative or synergistic effects, and any other effect which may pose an unreasonable risk of injury to health or the environment. Among characteristics to be considered are persistence (how long the material will still be active before it is degraded or otherwise converted to a biologically inactive form), acute toxicity (high dose over short time), subacute toxicity (produces effects but not fatal), and chronic toxicity (low-level doses over long periods) (Fig. 2.1).

Essentially, the law applies to all new chemical substances, and to new uses of existing substances. Notification and test data must be submitted 90 days before the manufacture or significant new use is begun. If the data are acceptable to the EPA, clearance will be given, otherwise the EPA may request additional evidence, or prohibit or limit production and use. The EPA will compile and keep current a list of chemical substances which it finds unacceptable.

Polychlorinated biphenyls (PCBs) are specifically considered in the Toxic Substances Control Act, with provision for new regulations for disposal of PCBs, and "clear and adequate" instructions in the marking or labeling. In 1 year (January 1, 1978), totally enclosed systems were required for PCBs, and after 2 years (January 1, 1979), no person could manufacture any PCBs or possess or distribute them after 2 and a half years (July 1, 1979), with certain provisions for exemptions. (Inasmuch as PCBs are the main dielectric insulating fluids used in electrical distribution transformers and capacitors, and all the substitutes to date have limitations, such as flammability under electric-arc conditions, the resolution of this substitution of PCBs presents an interesting and important technical and social, as well as environmental, development. See Chap. 8.)

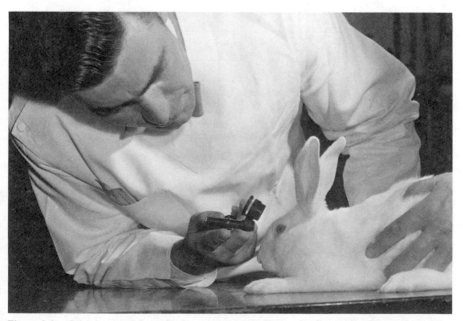

Figure 2.2.   Scientist examines rabbit for chloracne (see Chapter 8). Courtesy of Dow Chemical Co.

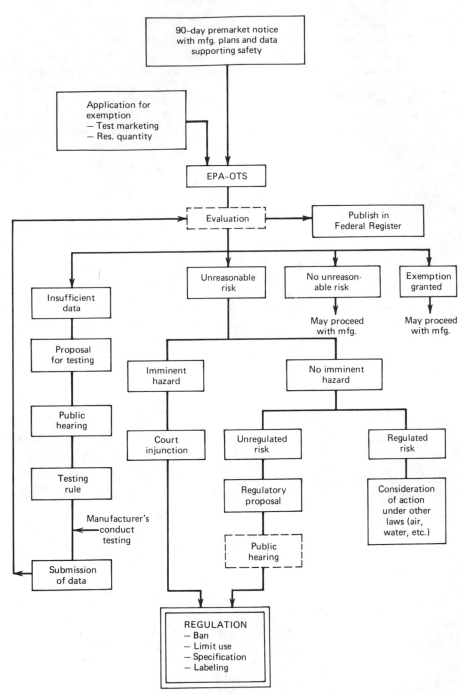

Figure 2.3. Regulatory scheme for new product under TSCA.

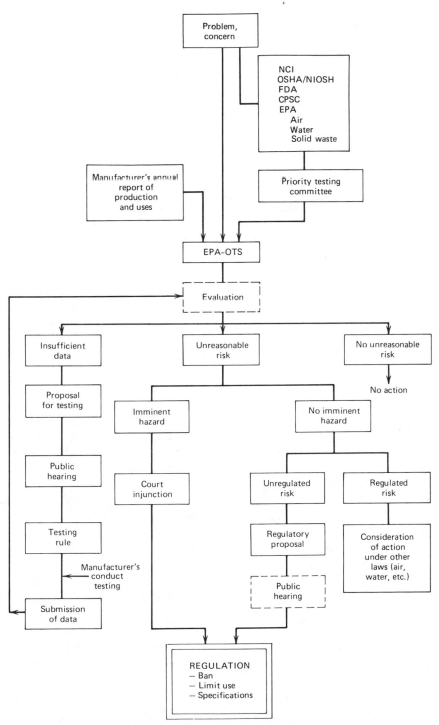

Figure 2.4. Regulatory scheme for existing product under TSCA.

Specifically exempt from the law are chemicals manufactured only in small quantities (as defined by rule) solely for purposes of scientific experimentation or analysis, or chemical research, analysis, or product development.

The bill provides for fines of up to $25,000 per day per violation, but considers the gravity and extent of the violation and ability to pay, as well as previous history of violations, in setting fines.

Criminal penalties may be imposed on persons who "knowingly or willfully violate any provision of the act which defines prohibited acts."

Figure 2.3 outlines our understanding of control schemes for existing chemicals which the EPA elects to review, while Fig. 2.4 outlines the procedures for new chemicals.

The TSCA Inventory Reporting Regulations were published in the *Federal Register,* Volume 42, Part VI, December 23, 1977. Reporting forms and the instruction manual can be obtained from all EPA Regional Offices and the EPA Headquarters' Industry Assistance Office, Washington, D.C. 20460.

While the legal aspects have received much attention in recent years, the long and sincere interest of major companies with the safe utilization of their products, both in manufacture and in the hands of the consumer, is often overlooked. The Dow Chemical Company established a toxicology laboratory in 1933. Much information was published in the open literature, including details of inhalation chambers useful for animal investigations (Fig. 2.3), which was described in an article by D. D. Irish and E. M. Adams, "Apparatus and Methods for Testing the Toxicity of Vapors," *Industrial Medicine* (Industrial Hygiene Section), January 1940, pages 1–5. Another example is illustrated in Fig. 2.2 which was originally published in Adams et al., "The Response of Rabbit Skin to Compounds Reported To Have Caused Acneform Dermatitis," *Industrial Medicine* (Industrial Hygiene Section), January 1941, pages 1–4 in which the rabbit ear test was used to detect the presence of chloracnegens long before the "dioxin" controversy. In November 1955, Dow established a separate biochemical research facility to coordinate all toxicity and biological activities. Other major companies, especially DuPont and Union Carbide, have been active in studies of human as well as animal effects from chemicals, and the companies should be consulted for information when questions arise about safe handling and use of their products.

The EPA has ordered an estimated 10,000 chemical makers to keep records of health-related complaints filed by employees. Congress ordered that these files be kept for at least 30 years. (See "EPA Issues Rules on Health Records of Chemical Firms," *Wall Street Journal,* August 23, 1983, p. 6.) (Rule is effective November 1983.)

BIBLIOGRAPHY

American Mutual Insurance Alliance, *Handbook of Organic Industrial Solvents, Technical Guide No. 6,* 3rd ed., rev. 1980. (Available from American Mutual Insurance Alliance, 20 N. Wacker Drive, Chicago, IL 60606.)

Borden, W. O., and Gibson, R. W., "Dangerous Materials and Related Problems," presented to California Fire Chiefs Association, University of California at Davis, March 18, 1969, Stanford Research Institute, Menlo Park, California.

Brown, J. M., "Legal-Social-Economic Considerations Affecting Hazardous Materials Decision Processes," Preprint 38A, presented at the Symposium on Legal, Social, and Personal Implications of Materials, Part I, Materials Conference, Philadelphia, Pennsylvania, March 31, 1968–April 4, 1968, American Institute of Chemical Engineers, New York.

Burgess, William A., *Recognition of Health Hazards in Industry: A Review of Materials and Processes,* Wiley–Interscience, New York, 1981.

Caglioti, L., *The Two Faces of Chemistry,* MIT Press, Cambridge, Mass., 1983.

Chen, E., *PBB: An American Tragedy,* Prentice-Hall, Englewood Cliffs, N.J., 1979.

Chissick, S. S. and Derricott, R., *Asbestos: Vol. 2: Properties, Applications and Hazards,* John Wiley & Sons, Chichester and New York, 1983.

"Deaths Due to Toluene Poisoning," *Occup. Hazards* **15**, 59–60 (January 1978).

deTreville, R. T. P., "The Role of Industrial Hygiene in the Proper Engineering of Materials," Preprint 43E, presented at the Symposium on Legal, Social and Personal Implications of Materials, Part II, Materials Conference, Philadelphia, Pennsylvania, March 31, 1968–April 4, 1968, American Institute of Chemical Engineers, New York.

"Formaldehyde: Evidence of Carcinogenicity," NIOSH Current Intelligence Bulletin 34, April 15, 1981. (Available from the National Institute for Occupational Safety and Health, Cincinnati, OH 45226.)

"ICWA Health and Safety News: OSHA Disallows Emergency Benzene Standard and Benzene—Cancer Risk Demands Controls," *The Chemical Worker,* Vol XLIII, No. 7 (August 1983), International Chemical Workers Union.

Katz, D. L., Woodworth, M., and Fawcett, H. H., "Factors Involved in Cargo Size Limitations," Report to the U.S. Coast Guard by the National Research Council Committee on Hazardous Materials, July 29, 1970, National Research Council, Washington, D.C.

Kraybill, H. F., and Mehlman, M. A., *Environmental Cancer, Advances in Modern Toxicology,* Vol. 3, Wiley, New York, 1977.

McCrone, W. C., "The Identification of Asbestos by Polarized Light Microscopy," 1982. (An audiovisual program—185 35-mm color slides with audiotape cassettes and program guide—available from F. I. Scott & Associates, P.O. Box 86, Check, VA 24072.)

Murphy, A. J., "Prevention Requires Teamwork," Preprint 43C, presented at the Symposium on Legal, Social, and Personal Implications of Materials, Part II, Materials Conference, Philadelphia, Pennsylvania, March 31, 1968–April 4, 1968, American Institute of Chemical Engineers, New York.

NTP Technical Bulletin, Issue No. 6, January 1982. (Available from National Toxicology Program, Department of Health and Human Services, Public Health Service, P.O. Box 12233, Research Triangle Park, NC 27709.) See also "Annual Plan, Fiscal Year 1982," NTP-81-94, National Toxicology Program, March 1982, same address as above.

"PCB's and the Environment," Interdepartmental Task Force on PCB's Report No. ITF-PCB-72-1, March 20, 1972. (Available from National Technical Information Service, Springfield, VA 22152.)

"PCB's in the United States—Industrial Use and Environmental Distribution, Task I," EPA Report No. 560/6-76-005, February 25, 1976, Final Report, Office of Toxic Substances, U.S. Environmental Protection Agency, Washington, D.C.

"Pipeline Accident Report, Mid America Pipeline System, Anhydrous Ammonia Leak, Conway, Kansas, December 6, 1973," National Transportation Safety Board, Report No. NTSB-PAR-74-6, Washington, D.C.

Rutter, M. and Jones, R. R., *Lead versus Health, Sources and Effects of Low Level Exposure,* John Wiley & Sons, Chichester and New York, 1983.

Safety Information Report No. SB 82-11/3365A, San Francisco, California, August 25, 1981, issued 1982 by the National Transportation Safety Board, Washington, D.C. 20594 ("PCB in Gas Pipelines").

Squire, R. A., "Ranking Animal Carcinogens: A Proposed Regulatory Approach," *Science,* **214,** No. 4523, 877–880 (November 20, 1981).

"The New Multinational Health Hazards," International Chemical Federation Meeting, Geneva, Switzerland, October 28, 1974–October 30, 1974.

Waldholz, M., "Potential of Costly New Superdrugs Leaves Doctors Excited but Wary," *Wall Street Journal,* March 11, 1982, p. 29.

"Yellow Rain: Gaining Speed," (Review and Outlook), *Wall Street Journal,* March 11, 1982, p. 26.

Occupational Exposures of Women

American Society of Anesthesiologists, "Effects of Trace Anesthetics on Health," NIOSH Contract HSM-99-73-003, October 16, 1972–October 15, 1978.

Anon. "Arsenic Standard Greatly Reduces Risk of Lung Cancer in Workers, OSHA Finds," *Wall Street Journal,* April 12, 1982, p. 14.

Brown, W., "NLRB Finds Safety Data Essential to Contract Talks," *Washington Post,* April 14, 1982, p. A21.

Fishbein, L. *Potential Industrial Carcinogens and Mutagens,* EPA 560/5-77-005, May 5, 1977, Office of Toxic Substances, Environmental Protection Agency, Washington, D.C.

Friedan, B., *The Second Stage,* reviewed in the *Wall Street Journal,* December 4, 1981, p. 34.

Hricko, A., with Brunt, M., *Working for Your Life; A Woman's Guide to Job Health Hazards,* Labor Occupational Health Program/Health Research Group, University of California, Berkeley, June 1976.

Hunt, V. R., *Occupational Health Problems of Pregnant Women,* A Report and Recommendations for the Office of the Secretary, Department of Health, Education, and Welfare, April 30, 1975. (Order No. SA-5304-75, The Pennsylvania State University, University Park, Pennsylvania.)

"Mutagens and Teratogens," subfile of "Registry of Toxic Effects of Chemical Substances." (Available from the National Institute for Occupational Safety and Health, Cincinnati, OH 45226.)

Nichols, E. E., *Guidelines of Pregnancy and Work,* NIOSH Contract No. 210-76-0159, American College of Obstetricians and Gynecologists, Chicago, Illinois, September 1977.

"NIH Guidelines for the Laboratory Use of Chemical Carcinogens," NIH Publication No. 81-2385, May 1982. (Available from Frederick Cancer Research Laboratory, Frederick, MD 21701.)

"Occupational Health, Part II, Occupational Health Hazards for Pregnant Women," Discussion Paper No. 14, December 1976. (Available from Working Women's Centre, 423 Little Collins Street, Melbourne, Australia.)

Proceedings, Conference on Women and the Workplace, June 17, 1976–June 19, 1976, Washington, D.C., Society for Occupational and Environmental Health, Washington, D.C.

Schechter, D., "Untangling the Asbestos Mess," *Occup. Health Safety* **51,** No. 2, 30–34(February 1982).

"The Dilemma of Regulating Reproductive Risks," *Business Week,* August 29, 1977.

"The Workplace," *The Spokeswoman,* 7–11 (July 15, 1976).

"Women in Employment," Vol. II, *Occupational Health and Safety,* International Labour Organization, Geneva, Switzerland, 1971, pp. 1501–1504.

Waldron, H., *Lecture Notes on Occupational Medicine,* 2nd ed. Blackwell Scientific Publications, Oxford, England, 1979.

Regulation as a Control For Low Risks

The tendency of society to attempt a solution of technical issues by legislation, and the regulation of chemical health and safety problems by governmental agencies, at the local, state, and federal level has been especially evident in the past 15 or so years. As noted in Fig. 1.2, the number of laws which pertain to toxicology and safety-related matters has increased from 3 to 13, not including laws establishing the National Fire Prevention and Control Administration in 1974, the Department of Transportation in 1968, the Transportation Safety Act in 1974, and others. The latter established the independence of the National Transportation Safety Board with overview of all modes of transportation. Every law has had sufficient justification to be passed by the Congress after much review and voluminous legislative history that extended over several years of hearings, and, at the time of signing by the President, was hailed as a historic landmark panacea to the problems it addressed. Unfortunately, when laws are interpreted by the courts at various levels, translated into regulations by the counts, and enforced by persons far removed from the spirit of the original issue, often with little interest or background in the problem, the interests of society may be less served than desired. The innate feeling that "government" can protect individuals from harm is less than a perfect concept in the real world. Decisions are made for one reason, and with justification, to be challenged by persons who feel they are not being adequately protected. The controversy over the spraying of 2,4,5-T in forests to reduce the undergrowth, frequently with emotional overtones which tend to obscure scientific inquiry, and the recent attempts to destroy marijuana plants on state and federal lands in Georgia and Kentucky by spraying with paraquat, a herbicide which, under some circumstances, can produce serious lung effects, are examples. Another longstanding question is the support of tobacco farmers by federal subsidy, while, at the same time, the Surgeon General continues to report that cigarette smoking is a major cause of lung problems, especially among the female population, where a 450% increase in lung cancer has been reported in the past 30 years, probably due to increased smoking by women under stress or peer pressure.

While RCRA (Resources Conservation and Recovery Acts) and SUPERFUND (Chap. 7) are the specific orientation of this book, the overall regulatory approach is of much concern, since it reflects on both our understanding of the specific problem, our economic assessment (or cost/benefit ratio), and the additional tasks on the already overburdened administrative and legal systems.

This approach by regulation becomes especially questionable when the risk is either low or relatively unknown. The control of alcohol and drugs by laws is, at best, a token of concern, even in occupations where sober concentration on the job is vital, as in the crew personnel on railroads transporting both people and hazardous cargoes.

Professor Lester Lave, writing in the *Wall Street Journal,* has noted that health and safety legislation sets unreasonable expectations of zero risk or zero discharge.[1] He notes that an area becomes a candidate for regulation as soon as scientists can identify a risk. This identification sets the regulatory machinery in motion toward a goal of zero risk, insofar as it is perceived by the regulators to be feasible. For major risks, such as lung cancer from asbestos and the much misunderstood regulation of vinyl chloride *monomer,* body counts can be used. Data were not collected systematically and science was less advanced, so a risk was recognized only when a major problem existed. Even today, years after the Supreme Court rejected the reduced regulation of the permissible exposure limit of benzene from 10 to 1 ppm because the OSHA had failed to sufficiently show "significant risk" to justify the change, a twilight zone of exposures is being condoned, and other substances, such as ethylene oxide and ethylene dibromide, are not on firm grounds in spite of extensive studies.

Professor Lave notes that producers of toxic chemicals have faced sensitive risk detection for more than three decades. Toxicologists have been perfecting highly sensitive tests for carcinogenicity and for reproductive risks, not just for the one-time acute exposure. As the Professor notes, "the chance of adverse effects occurring with the use of most chemicals is miniscule." A zero-risk goal is not realistic; it would paralyze the entire economy. Some hope is in recent actions where the Nuclear Regulatory Commission set a risk goal for nuclear power plants—that risks to the surrounding population may not be increased more than 0.1%, one part in 1000. The Food and Drug Administration now acts as if a food additive is not really a carcinogen if it would cause less than one cancer per million lifetimes of those exposed. The Environmental Protection Agency regards one cancer in 100,000 lifetimes as negligible.

Congress has faced up to the problem of regulating pesticides by the EPA and drugs by the FDA, instructing the agencies to balance risks against benefits.

W. W. Allison, an astute observer of the scene, has noted that the media, some academics, and legislatures and others have given headline-seeking knee-jerk responses to both hazards and alleged hazards. The police-citation approach which existed at the OSHA for 10 years did not improve accident prevention any better, if as well, as recent movements towards consultations and understandings. The answer, Allison feels, is more objective academic, government, and industry research and support in finding the *real* root causes and *right* solutions to chemical hazards. This requires a conciliatory and cooperative approach by government, academics, and industry similar to the approach of the Japanese government towards its industrial sector. Companies with the right solutions must help others understand and emulate them with the support of the bureaucracy.[2]

An example of cooperation among nations to provide a scientific basis for possible controls on acidity from industrial sources was announced in the *Wall Street Journal,* August 24, 1983, page 46. The U.S. and Canada Sign Accord for Joint Study of the Acid Rain Problem, in which an inert tracer gas track upper air movements, hopefully will provide sound basis for further discussion of the "acid-deposition" question.

An interesting example recently of the "toxicity versus hazard" question is noted by Phil Wingate in the EPA dilemma to regulate arsenic emissions from a smelter in Washington State. If the EPA sets too high a figure for emission control, it may harm the health of workers and residents of the community. If it sets too low a figure, it may close the plant, resulting in unemployment. A zero figure would close the plant immediately. As Wingate correctly notes, no one really knows how low the exposure to arsenic must be kept before it is harmless to people. Years ago we learned of the arsenic eaters of the Andes mountains, where acclimation to the low oxygen tension of high altitudes has improved by careful and measured introduction of arsenic in the diet to increase the number and vitality of red blood cells.

It is true that arsenic is widely dispersed in nature. Wingate has calculated that the oceans, in addition to tons of moderately toxic sodium chloride (salt), contain 3 billion pounds of arsenic. Yet, excessive doses of arsenic are doubtlessly fatal, as has been demonstrated for years, as are excessive amounts of salt. The "threshold" of risk/benefit probably varies for different persons, and is not presently known with any absolute certainty. A threshold value for ambient air in the vicinity of a smelter may differ significantly from a TLV for occupational exposures, for 8-hour days.

Wingate also cites another example, chloroform, which was used in massive exposures from 1875 to 1925 as an anesthetic. It is now known that it is hazardous to the liver and is no longer used as an anesthetic. Safer agents have been developed.

As Wingate concludes, while a little bit of many things (including arsenic) can be tolerated, too much of even a good thing tends to become bad.[3] Our problem is to define the "little bit."

REFERENCES

1. L. Lave, "The High Cost of Regulating Low Risks," *Wall Street Journal,* August 19, 1983, p. 18.

2. W. W. Allison, "Reaction to Hazards," *Occup. Health Safety,* **52,** No. 8, 44 (August 1983).

3. P. J. Wingate, "Everything in Moderation—Arsenic Included," *Wall Street Journal,* September 8, 1983, p. 30.

4. Third Annual Report on Carcinogens, September, 1983, National Toxicology Program Information Office, Box 12233, Research Triangle Park, N.C. 27709.

5. "Melanoma Risk and Socio-Economic Class," *Science News,* **124,** 232 (October 8, 1983).

6. B. N. Ames, "Peanut Butter, Parsley, Pepper and Other Carcinogens," *Wall Street Journal,* February 14, 1984. "The EDB Flakeout, Review and Outlook," *Wall Street Journal,* February 14, 1984, p. 32.

7. C. Peterson, "The Hyperbole War Over EDB," *Washington Post,* February 12, 1984, p. 35.

3
Toxicity—Part 2, Toxicity Versus Hazard

3.1. HOW TOXIC—HOW HAZARDOUS?

These two related thoughts, toxicity and hazard, are frequently combined into one compound question, yet in their practical aspects they may require entirely different answers. This discussion will point out some of the factors which must be considered in evaluating the true meaning of the "hazard" of a substance suspected of being "toxic." To some people, the Toxic Substances Control Act (P.L. 94-469) was a new thought and an entirely new approach to the control of potentially hazardous materials. Some authorities feel it was the most important piece of legislation pertaining to the environment to be enacted by the Congress. The act, which became generally effective January 1, 1977, gave the EPA's administrator broad authority to compile inventories of existing substances; to require the chemical industry to conduct extensive testing of substances; to delay the manufacture and marketing of a new product if questions arise as to whether or not it is safe for intended use; to bar or place restrictions on the marketing of existing new substances or of new applications for the substances; and, to require the maintenance of such records as the administrator may reasonably require. However, this act, which placed the word "toxic" in a specific context, probably created as much confusion and legal dispute as any previous legislation. It was not the first attempt to control "toxic" substances by law (see Fig. 1.2). The word "toxic" has been loosely applied to many different effects, when actually it should be related to general, systemic effects of a substance in living animals or in human beings. Almost every substance will produce injurious effects to some degree in a living body, because even safe substances, or those that at various times have been officially classified as GRAS (generally regarded as safe) in excessive amounts or in certain dosages can produce injurious effects. Substances such as salt, for example, which is by any definition a moderately toxic substance [M. Allen, "A Time for Spices (as Replacement for Salt in Foods), *Saturday Evening Post,* Vol. 255, No. 6, September 1983, pp. 90–92.], baking powder, commonly found in practically every baked product, and sugar, which certainly in excessive amounts can cause very serious dysfunctions and lack of coordination in the body's system, are all potentially toxic materials, if by toxicity we mean an adverse effect on the human body. The key questions, of equal im portance, are:

How much of a substance is needed to produce a toxic effect?

How likely is this amount of material to enter the body where it can actually produce this effect?

Substances differ widely in their relative toxicity, in their ability to enter the body, and in the effects they produce.

The word "toxicity" has many definitions. We prefer the discussion be limited to a very simple definition for this purpose, namely, that toxicity is the effect produced by an excessive amount of a substance being incorporated into the body system beyond that which the body can eliminate or tolerate without injury. *Toxicity is not a specific physical constant determined by standardized devices,* such as used to determine specific gravity or melting point. While there are protocols for determining the toxic effects of materials, most of these protocols require administering large quantities of substances to animals, since they require excessive doses of a substance to produce an obvious effect, and, since they are also subject to interpretation, the absolute standardization of materials with respect to toxicity probably is beyond our present ability. This in no way means we should ignore toxicity, because toxicity is a very serious effect produced by many materials, and not so seriously considered with respect to others. Our knowledge is less than complete, and with due respect to the wisdom of Congress in passing Toxic Substances Control Act, we know that the ability to specifically identify, quantify, and make proper use of toxicity information, is a long series of evolutions which probably will require many years, if in fact absolutes are ever obtained in the biological systems involved. It cannot be stressed too strongly that toxicity is not a property of the substance itself, but, rather the degree to which the substance affects living cells. We stress the word "affects" and the word "living." No meaningful definition of toxicity can ignore these fundamentals.

Even standardized batches of laboratory animals where hundreds or thousands of rats, mice, guinea pigs, hamsters, monkeys, or other animals are bred under laboratory-type conditions, will exhibit considerable difference in their characteristics. In one study at the University of Texas some years ago, white rats were studied with respect to differences in individual characteristics. One rat drank 15 times more alcohol than another, and another rat traveled 6 miles while his less active sibling was moving 150 ft. If rats, bred and carefully selected for uniformity, are such nonconformists, consider how much more human beings differ from each other and from the standard curve. The problem of individual variation has been demonstrated many times in the field of toxicology, as well as in other phases of life. The cliché, "one man's meat is another man's poison," is appropriate in summing up man's experience in this respect. Just as there is no "standard animal," there is no standard human. Both animals and humans are subject to wide variations.

Toxicity cannot be measured at all or even satisfactorily quantified until a definite recognized change has occurred in an animal or a human. These changes may be very small, easily overlooked, and quite subtle. Such changes as impaired judgment and delayed reaction time may be involved at levels too low for the production of body damage.[1] Nevertheless, effects can exist and depending on the length of the experiment, the number of animals used, and the sophistication of the pathologist who makes the final judgment, the results can either be meaningful or not meaningful. It is not unusual, for example, in the studies which have been made over the years, to find a chemical listed as producing carcinogenic effects in one species of animal, but not in another species of animal. Laboratory animal specialists have pointed out repeatedly that there is a great variation even among various strains of rats and mice with respect to the susceptibility to various changes, that is, acute toxicity or carcinogenic, mutagenic, teratogenic response to given doses. Until we fully appreciate that no animal experiment can approach the precision of computers, mathematical models, and other sophistication which has been introduced into the system, can we approximate the knowledge gained from human observation and human exposure.

Animals may react in a much different manner than humans to the same exposures. Mules, for example, do not develop silicosis, a disease of the lungs, even while working beside miners who develop the disease. Crystalline penicillin-G is essentially nontoxic in animals, with the exception of guinea pigs. Guinea pigs are particularly sensitive to penicillin and to certain other antibiotics. Doses as small as 7000 units per kilogram may produce serious effects and finally death within a few days. Doses as high as 9.8 million units or 5.93 g/kg were tolerated in mice, while humans can tolerate tremendous doses. Humans can tolerate doses on a daily basis as high as 86 million units for a 28-day period, if they are perfectly normal and well; but, those that are allergic will find that even small doses produce severe and, occasionally, fatal shock. This is why penicillin is no longer the "wonder" drug that it was widely hailed to be 15 or 20 years ago. Animals may survive a relatively large or acute dose, but die from smaller doses over a long period of time because the ability of the body to tolerate by various modes of metabolism or elimination is not sufficiently understood.

To understand toxicity, as related to hazard, we must examine certain fundamental "real-world" areas.

3.2. COMPOSITION OF THE SUBSTANCE AS ACTUALLY USED OR HANDLED

It is futile to attempt any evaluation of hazard without specific and definite information on composition. To say a solvent mixture "contains mineral spirits" is to present inadequate information, since mineral spirits vary widely in composition,

TABLE 3.1. Major Laws Controlling Toxic Substances

Environmental Protection Agency

Toxic Substances Control Act of 1976—places heavy reporting burden on industry; the EPA can demand premarket testing of some chemicals.

Safe Drinking Water Act of 1975—carcinogens and toxic substances in public drinking water supplies.

Resource Conservation and Recovery Act of 1976—disposal of toxic and other hazardous wastes in landfills, by incineration, and so forth.

Water Pollution Control Act Amendments of 1972 and 1977—discharges of hazardous effluents into the nation's waterways and the ocean landward of the 3-mile limit; cleanup of hazardous spills on land and in water.

Marine Protection, Research, and Sanctuaries Act of 1972—ocean dumping from 3 to 12 miles at sea (enactment of 200-mile limit in spring 1977 may extend this).

Clean Air Act Amendments of 1970—allows the EPA to set national emission standards for hazardous air pollutants; standards now exist for beryllium, asbestos, mercury, and vinyl chloride.

Federal Insecticide, Fungicide, and Rodenticide Act of 1972—covers the sale and use of economic poisons, and foodstuff treated with such poisons; counterpart legislation exists for enforcement by the USDA and the FDA.

Food and Drug Administration

Federal Food, Drug, and Cosmetic Act of 1906 (amended 1938 and 1962)—bars any detectable amounts of carcinogens in foods and cosmetics; specifically exempts hair dyes from enforcement provisions.

Occupational Safety and Health Administration

Occupational Safety and Health Act of 1970—hazardous materials in the workplace, or in products bought for use in the workplace.

Consumer Product Safety Commission

Consumer Product Safety Act of 1972—specifically excludes tobacco, foods, drugs, and cosmetics from regulations; created commission to enforce, among other things, earlier statutes.

Federal Hazardous Substances Act of 1927 (amended 1976)—flammable, corrosive, allergenic, or toxic materials in consumer products.

Flammable Fabrics Act of 1953 (as amended)—clothing as well as fabrics used for other purposes in the home.

Poison Prevention Packaging Act—the law that produced childproof caps.

Department of Transportation

Oil Pollution Act of 1961—oil spills and chemical spills from ships.

Dangerous Cargo Act, Tank Vessel Act, Ports and Waterways Safety Act 1972 and Pipeline Safety Act—various railroad and truck transportation safety laws.

and also the percentage of mineral spirits in the mixture may be very small or quite high. Other constituents in the mixture may be far more hazardous, such as benzene or carbon tetrachloride. If a proprietary-brand solvent or mixture is involved, the maker will usually reveal the complete formula on a confidential basis to a responsible person for the use, if given assurance that the information will not be used against the supplier's best interests or passed on to anyone else. The subject of confidential information has received much attention, especially with reference to release of information to government agencies, such as the EPA, the FDA, the NIOSH, the OSHA, and the CPSC. Since the matter is largely legal, rather than technical, the advice of a legal expert or counsel should be sought. It should be noted that the composition of "trade-name" materials may change from time to time, with little or no change in the name or label.

If the supplier will not cooperate, two alternatives are always suggested (1) analyze the substance or (2) locate a more cooperative supplier. Usually the supplier will cooperate if the supplier realizes his or her business depends on cooperation, and this approach is usually faster, more accurate, and more economical for all concerned than analysis. With the wide use of techniques such as chromatography and infrared spectography, however, analyses are much easier to obtain than previously, and the complete analysis should be obtained if we are suspicious a hazard may be involved.

3.3. TOXICITY IN ANIMALS

Once we have learned what is involved chemically, we can turn to the literature in the hope that our substance has been investigated and the data published. At this point extreme care must be exercised, since toxicity values are by no means absolute; they can be considered only as yardsticks of activity. Spector[1] lists five conditions that influence the toxicity of any given substance:

Dose.   Generally the larger the dose, the more rapid the action.

Rate of Absorption.   The faster the rate of absorption, the quicker the action. Food or oils in the stomach slow this action. For skin absorption, the larger the area involved and the longer it is in contact, the faster the effect.

Route of Administration.   Toxicity is greatest by the route that carries the toxic substance to the bloodstream most rapidly. In decreasing order of speed, routes for most substances are:

Intravenous (into a vein)

Inhalation (breathing)

Intraperitoneal (into the abdominal cavity)

Intramuscular (into a muscle)

Subcutaneous (under the skin)

Oral (by mouth)

Cutaneous (on the skin)

Site of Injection.   With subcutaneous injections, toxicity may be affected by the density of the subcutaneous tissue. In intravenous administration, the rate of injection, or the amount of toxic material injected per minute, will considerably influence the value of the toxic dose.

Other Influences.   Disease, environmental temperature, habit and tolerance, idiosyncrasy, diet, and season of the year may all affect toxicity. The toxicity of chemicals will also vary with the species of animals used and sometimes with different strains of the same species. Within the same strain, the toxicity may differ with age, weight, sex, and the general conditions of the animals. The time to produce death, or the period of time for which fatalities are counted, may also be a factor.

There are several units in which the toxicity dose is expressed. The most frequently used are: LD or lethal dose (the amount which kills an animal), the MLD or minimum lethal dose (the smallest of several doses which kills one of a group of test animals), LD_{50} or lethal dose for 50% (the amount which kills 50% of a group of test animals, usually 10 or more), and LD_{100} or lethal dose for 100% (the amount which kills 100% of a group of test animals, usually 10 or more). Sometimes D is replaced by C in the above symbols, and the work "concentration" used instead of dose, as LC = lethal concentration, when referring to vapor concentration in air.

The usual form in which lethal doses of solids and liquids are expressed is in milligrams of substance per kilogram of body weight of the animal, or abbreviated as mg/kg or g/kilo. Since a 150-lb adult weighs about 70 kg, it might be expected that animal data could be translated into human data by multiplying the mgm/kg dose by 70. This practice is filled with pitfalls, and it should be used only as a "degree-of-magnitude" rough calculation. For reasons already mentioned, plus the important fact that there is no laboratory animal (except possibly the higher apes) which reacts to chemicals as does man, extreme care should be used in applying animal data to humans.

3.4. HUMAN EXPOSURES

The real value and ultimate test of toxicity data, of course, is what actually happens to humans. Here we see even clearer that toxicity/hazard is not a simple matter. Some substances highly hazardous to young children (up to age 4 years) are relatively nontoxic to adults, because no adult would knowingly eat or drink them except by the highly unusual accident or a suicide measure.[2]

Two common materials which cause serious poisonings in children are kerosene and aspirin. By chewing on cribs, windowsills, and toys painted with lead-containing paint, children may be poisoned with lead. The widespread presence of lead in water, in air, exhaust fumes, agricultural products grown along highways, and in the food consumed in major cities gives this topic high priority. (See L. Caglioti, *The Two Faces of Chemistry*, MIT Press, Cambridge, Mass., 1983, pp. 91–93.)

A highly unusual case, reported a few years ago, developed from a teenage boy chewing on a lead "sinker" which he used in fishing. Swans on the Thames near London have been seriously affected by lead from fishing sinkers. Medications, especially aspirin and sleeping pills, are often eaten by young toddlers, occasionlly with fatal results. The operation of over 500 poison-control centers in the United States and Canada to advise the physician on an emergency basis of the composition and recommended treatment in poisoning cases has helped to make the public more aware of accidental poisoning cases and hence to prevent accidents involving drugs and other chemicals. In general, these are coordinated by state health offices in larger hospitals. A national clearinghouse operated by the Department of Health and Human Services coordinates the information base.[3]

Another practical problem encountered in evaluation of hazards is that much necessary data can only be obtained by experience. For example, some substances such as the isocyanates have a very low level of true toxicity, but in extremely small concentrations in air they can cause bronchial irritation from sensitization in sensitive poisons. The irritating aspects of materials are not adequately reflected by animals. A mixture of diphenyl and diphenyl ether is a widely used heat-transfer agent having a very low-toxicity rating, but when accidentally sprayed in a face it may cause serious, almost fatal, respiratory effects due to irritation. Hydrogen sulfide is an example of a gas which in higher concentrations soon paralyzes the nose so the odor cannot be used as even a rough estimate of concentration for this highly dangerous substance. Irritation by a substance varies with people. Some people frankly state they like the odor of low concentrations of pyridine and of mercaptan, while others are irritated, annoyed, and affected by the same concentration.[4] Boric acid is an example of a chemical which has been used in the treatment of burns for many decades, but whose toxicity has been recently recognized as too high to justify its use in this application. Another important drug, digitalis, has been successfully administered for over 200 years for treating congestive heart failure, as well as atrial arrhythmia. However, nearly a third of the patients on this drug develop toxicity, and many die if the drug is improperly used. The dosage is very critical since digitalis (or digoxin) has a half-life in the body of 36 hr, indicating that half a given dose will be excreted in 36 hr.[5]

Beyond the gross dosage problems, as reflected in poisonings, there are the practical day-to-day exposures encountered in industry. This is the real practical test of toxicity. What does the substance, either alone or in combination with other substances, do when breathed or absorbed in other ways at rates which will probably vary over wide limits during the day for several hours a day for a certain period of time, such as a 40-year working lifetime? To guide the control of such exposures, Professor Warren Cook tabulated and published in 1945 recommendations of maximum concentrations permissible for many common substances. Since 1947, the American Conference of Governmental Industrial Hygienists (ACGIH) has sponsored a committee which publishes an annual revision to the threshold-limit values, formerly referred to as MAC or maximum acceptable concentration values. Nearly 600 substances are currently listed by the committee. This list is not an official "standard," as such, but many states have adopted it as the working limits for their labor or health codes, and the 1969 edition became the official OSHA standard for most values.

Before referring to this list, however, it is wise to carefully read the preamble in order to understand exactly what the list is as well as what it is not. These values are not a measure of *relative* hazard. The early values were established largely on the basis of safe level to prevent damage from chronic exposures. As more data became available based on actual experience, many values were lowered to reflect irritation and other transitory or acute changes. Recently, subacute levels have been introduced to reflect comfort levels. In addition to the threshold-limit values, when evaluating hazards, we must consider the vapor pressure of the material which will determine the potential of attaining the exposure level under given conditions of use, such as ceiling values and time-weighted average values (TWAs), as well as other important factors which can relate to actual conditions of use. In recent years, short-term exposure limits (STEL) have been listed for many substances. Physical agents, such as noise, light, ionizing, and nonionizing radiation, and lasers have been considered by the ACGIH committees, in addition to chemical exposures and threshold limits established for them as well.

The United States is not the only country which has attempted to develop standards for exposure levels. While several countries use the basic data of the United States,[5,6] the Soviet Union has developed standards which are often considerably different than American standards. Apparently, the Russian values are based on behavioral toxicity studies, and hence reflect a data base different from ours.[7] A more complete understanding of what represents the earliest manifestation of injurious effect would be highly valuable.

Another source of information which is often helpful in evaluating the hazards of a substance is the manufacturer—a person who is truly interested in seeing that the substance is used without adverse effects. Both in published data sheets and in answers to specific inquiry by telephone or mail, manufacturers will usually give practical recommendations as to the precautions they believe necessary. The more specific the inquiry, the more helpful will be the reply. Regardless of the quantity (a few grams may represent more hazard than a million gallons), the fundamentals are the same and manufacturers will usually supply the information if requested and assured the inquiry is genuine.

Another guide to health safety information is open literature. Much is available for those who seek it. A two-part article in *Industrial and Engineering Chemistry* points out several excellent sources of safety information not widely used.[8] The National Safety Council[9] publishes information on many chemicals—information that helps in evaluating potential hazards. *Hygienic Guides,* published by the American Industrial Hygiene Association,[10] now cover nearly 200 substances in considerable detail. The National Center for Toxicological Information[11] located in Oak Ridge, Tennessee, is another national resource, as is the National Institute for Occupational Safety and Health[12] located in Cincinnati, Ohio. Several companies have made their product data sheets available. (Data on several hundred chemicals are tabulated in *Safety and Accident Prevention in Chemical Operations,* 2nd ed. H. H. Fawcett and W. S. Wood, Eds., Wiley–Interscience, New York, 1982, Appendix 1, pp. 865–875.)

3.5. CLASS OF TOXICITY INCLUDING THE SUBSTANCE

Ultimately we must decide on the degree of hazard presented by a substance. Classification of hazards goes far beyond toxicity, as illustrated by the excellent work of a National Fire Protection Association (NFPA) Committee headed by the late James J. Duggan which classified and labeled substances, especially larger amounts of substances in barrels or storage tanks, for emergency control purposes. The NFPA Standard 704-M, which resulted from this work, includes toxicity (hazard to life), flammability (fire hazard), explosion hazard (probability of explosion), and chemical reactivity (possible reactions with other nearby substances, if spilled or ruptured), which, combined with other essential information would be most valuable to emergency personnel, and its wide application should be encouraged.[13]

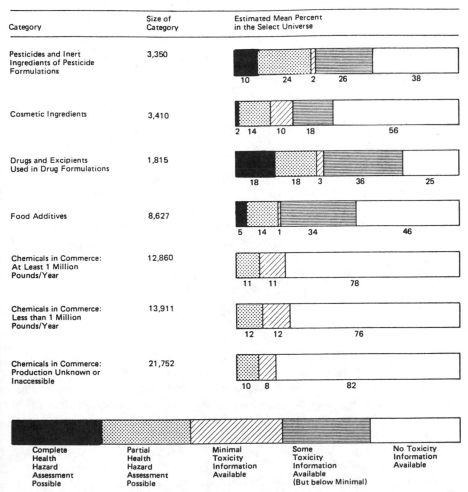

Figure 3.1 Ability to conduct health-hazard assessment. Reproduced from *Toxicity Testing: Strategies to Determine Needs and Priorities,* with the permission of the National Academy Press, Washington, D.C. 20418.

Hodge and Sterner tabulate toxicity into six classes[14]:

Toxicity Rating or Class	Probable Oral Lethal Dose (Human)	
	Dose	For a 70-kg Person (150 lb)
6—Super toxic	Less than 5 mg/kg	A taste (less than seven drops)
5—Extremely toxic	5–50 mg/kg	Between seven drops and 1 tsp
4—Very toxic	50–500 mg/kg	Between 1 tsp and 1 oz
3—Moderately toxic	0.5–5 g/kg	Between 1 oz and 1 pt (or 1 lb)
2—Slightly toxic	5–15 g/kg	Between 1 pt and 1 qt
1—Practically nontoxic	Above 15 g/kg	More than 1 qt (2.2 lb)

The National Academy of Sciences—The National Research Council has determined that the adequacy of toxicity data is far less than generally assumed (see Fig. 3.1.).

3.6. LIKELIHOOD OF RECEIVING A HARMFUL CONCENTRATION

The old cliché of the sea states, "Not all the waters of the seven seas can sink a boat until it gets inside the boat." Toxicity is much the same. A chemical in a bottle, in a tank, or in a boat is harmless as long as it is fully contained. In evaluating the hazard, therefore, the basic consideration should be based on how much of the substance will be in the air, or where it will be so it may be ingested or absorbed through the skin or the eyes. Once this is established, we can add the toxicity data, expressed as the dose or irritation or maximum allowable concentration or threshold-limit value, and find some measure for the actual hazard.

About 20 years ago, a relatively new chlorinated hydrocarbon was introduced on the market and widely promoted as a substitute for other solvents such as carbon tetrachloride. (Carbon tetrachloride had been recognized as too hazardous for "bucket" or "open container" operations, in spite of its relative safety insofar as flammability is concerned.) The new substance had many characteristics and properties of carbon tetrachloride—its vapor pressure, cleaning ability, and ability to dry without residue were similar to carbon tetrachloride. Extensive animal exposure established a firm basis for assigning this substance a threshold-limit value of 500 ppm for occupational exposures in contrast to 10 ppm then in effect for carbon tetrachloride. This limit has since been lowered, but still the consensus is that the potential hazard is much less than carbon tetrachloride for a comparable exposure. Although we agree this solvent is much safer than carbon tetrachloride and we continue to recommend it for many applications, we must point out that at least three fatalities have resulted from its improper use. Such solvent fatalities usually involve a careless use in a confined unventilated space. In one incident, a worker, while working at the bottom of the outer shell of a vacuum furnace, 38 in. in diameter by 49 in. deep, was cleaning oil and grease from the interior surfaces with steel wool and rags saturated with the solvent. The tank which was located in an open room had one 15-in.- and three 6-in.-diam. ports 28 in. from the bottom. The solvent was being used from an open coffee can. Approximately a quart had been used when the technician was found dead, about 50 min after the supervisor had last checked the work. Such cases do not indicate the solvent is not relatively safe—it simply underscores again that any solvent must be used with respect. *Use of any solvent in a confined space without adequate ventilation, especially by a person working alone or with only nominal, occasional supervision is extremely unwise.*[15a] In this connection, an advertisement used at one time by some distributors that a solvent is "20 times safer than carbon tetrachloride" is highly misleading, and represents a use of threshold limit values in an entirely different manner than intended. (The chemical reaction between this solvent and an aluminum hand-pump created an overpressure in a New York 12-story building, causing closing of a major avenue to rush-hour traffic. Normally the solvent is inhibited from aluminum reaction.[15b])

3.7. EMOTIONAL APPROACH TO THE SUBSTANCE

A few substances have come to be associated with high hazard in the public mind. If we consider "poisons" as synonymous with "hazards," we probably think of cyanide, lead, silica, arsenic, asbestos, and carbon monoxide. Snake venom, curare, benzene, and carbon tetrachloride might be included on second thought. The strange enigma of the lung disease related to beryllium and beryllium compounds and the reluctance of some technical persons to accept the facts about the substances, indicate that publishing data alone does not ensure that everyone will be informed or will believe. Some of the older tonnage chemicals, such as aniline, nitrobenzene, and hydrogen sulfide, are finally being recognized as hazardous, and there are many substances about which we know so little that it is impossible to even guess how safe or how hazardous they are in the intended applications. Criteria documents, as developed by the NIOSH for consideration by the OSHA in a standard setting, make a sincere attempt to evaluate hazard, but in some cases limited data have lead to emphasis which may not coincide with practices and procedures based on years of "real-world" experience. A western university reported the death of three students who were working with several bicycloheptadiene derivatives, previously not considered hazardous. Animal investigations are underway to elucidate the hazard.[16] Magic methyl, or methylfluorosulfonate, caused a fatality in the Netherlands before its respiratory hazards were recognized.[17] The need for new data even on the more familiar substances must not be overlooked. [The glycol ethers, such as 2-methoxyethanol (2ME) and 2-ethoxyethanol (2EE), have been recently reported as having the potential to cause adverse reproductive effects in male and female workers.[18] Gasoline, an even more common substance, has recently been questioned. The

American Petroleum Institute and a major oil company (Exxon) recently issued information to consumers on the health effects of gasoline. The use of gasoline in normal situations (such as pumping gas into a car or fueling a lawn mower) is highly unlikely to produce any of the symptoms listed below. However, inhalation of gasoline vapors can cause dizziness and irritation of eyes, nose, and throat. Prolonged breathing of high concentrations of vapors can cause headache, nausea, slurred speech, and difficulty in swallowing. Extremely high concentrations of gasoline vapors can lead to unconsciousness and even death. Gasoline absorbed through the skin can cause problems similar to those caused by breathing gasoline vapors. In studies with laboratory animals, gasoline vapors caused kidney damage and kidney cancer in rats, and liver cancer in mice. (Exxon, August 1983.)]

Extreme fear and anxiety about hazards may actually create accident situations, just as may apathy and ignorance. Personnel may be so fearful that they will become nervous and supersensitive—perhaps even allergic. The reverse condition, lack of adequate and proper respect for a hazard, may also contribute to accidents by encouraging carelessness and lack of protective measures. Exactly how to present the degree of hazard to personnel in their specific use of the toxic or corrosive material, so they will actually react with respect and confidence but not fear, is one of the challenges of supervision. In addition, the emotional state of the workers, their emotional stability and adequate adjustment to their jobs, their bosses, their company, and their home life, all may be far more important in evaluating actual on-the-job hazards than toxicity data alone. In all cases, it must be *both* the worker and the chemical which are the center of our attention (see Chap. 2).

The increased attention which the news media, especially television and radio, have given chemical emergencies has produced an awareness of the disadvantages of chemicals when they are misused or carelessly handled. Kepone, polychloronated biphenyl, polybromobiphenyl, and dibromochloropropane are chemicals which have received unfavorable publicity due to careless exposures or accidents which dilute the real benefits from their safe and proper use. The interface between emotions, science, and politics is clearly shown by Lois Gibbs at Love Canal[19] and by the novel *Fever,* concerning some bizarre exposures of benzene to a 12-year-old.[20] A chemical emergency is news but the day-by-day safe use of the substance is not.

The emotions play a significant role in influencing the impressions which society has of chemicals, the chemical industry, and chemists. The risk/benefit factor which may minimize the hazard of a substance is irritation ability. If a gas or vapor affects the upper respiratory tract sufficiently to cause sneezing, coughing, or extreme discomfort to the eyes or throat, the normal reaction is one of escape or repulsion, which tends to decrease the exposure. On the other hand, if the substance is not sufficiently irritating or objectionable, the practical danger is much greater, since the warning will be less or inadequate. This is one factor why carbon monoxide remains the serious hazard in both vehicle and industrial exposures—virtually no warning of serious exposure occurs before being overcome by the gas.[21] In a recent case, a 32-year-old woman attempted suicide by deliberate exposure to gas high in carbon monoxide. She survived 5 days. Tomographic studies confirmed the normal pathology findings of CO poisoning.[22] On the other hand, formaldehyde, which is highly irritating in low concentrations, was banned from home insulation of the UF-type (urea-formaldehyde) by the Consumer Products Safety Commission, largely because the irritation was easily demonstrated. (On August 26, 1983, the Justice Department announced it would not appeal a Supreme Court ruling that negated the ban on UF-type insulation. However, the economic impact of this ban continues. According to the *Wall Street Journal,* September 7, 1983, page 33, the presence of urea-formaldehyde insulation reduces a home's value by an average of 14%. . . . Although it is cost-effective insulation, urea formaldehyde gives off fumes that some say cause nausea, headaches, and more serious illnesses. Until a recent court action, the government had banned its use in homes. Nevertheless, an estimated 500,000 homeowners had it installed before questions were raised. Now, appraisers are hard pressed to set values for homes with urea-formaldehyde insulation. A Rochester, Wisconsin relocation consultant surveyed 98 appraisers on the subject and found that they slashed home values 5–40% from what the homes otherwise would sell for. Several appraisers said they base the discount on the cost of removing the insulation. Even when the urea formaldehyde has been removed, people are reluctant to buy the house, according to the consultant.)

3.8. SUMMARY

In summary:

1. Chemicals, per se, are not toxic or nontoxic (common usage meaning the ability of excessive amounts to produce damage to life).
2. Toxicity refers to the *effect on living cells* (usually animals or humans). Toxicty data must be carefully screened and examined before extrapolating into human experience.

Chemicals, even "extremely toxic," may or may not be hazardous depending on their use. *All chemicals can be handled and disposed of safely.* Until they reach and produce effects inside the body, toxic substances are not harmful. Toxicity then is a phenomenon of living organisms and is *not* a fundamental property of the contained material.

REFERENCES

1. W. Spector, *Handbook of Toxicology* (in two volumes), Saunders, Philadelphia, 1956–1957.
2. Sir Michael Foster (1836–1907), *J. Am. Med. Assoc.* (Editorial), **186,** 1167–1168 (December 28, 1963).
3. National Clearinghouse for Poison Control Centers, Rockville, MD 20857.
4. "Threshold Limit Values" (updated annually). (Available from the American Conference of Governmental Industrial Hygienists, 6500 Glenway Avenue, Bldg. D-5, Cincinnati, OH 45211.)

5. J. E. Meissner, and L. N. Gever, "Digitalis—Reducing the Risks of Toxicity," *Nursing 80,* **10,** No. 9, 32–38 (September 1980) (7 refs.).

6. M. A. Winell, "An International Companion of Hygiene Standards for Chemists in the Work Environment," *AMBIO* **4,** No. 1, 34–36 (1975).

7. G. J. Ekel and W. H. Teichner, "An Analysis and Critique of Behavioral Toxicology in the U.S.S.R.," Contract No. HSM-99-73-60, December 1976, National Institute for Occupational Safety and Health, Cincinnati, Ohio.

8. H. H. Fawcett, "Who Knows What About Chemical Safety," *Ind. Eng. Chem.* **52,** No. 8, 85A–88A (June 1960); **52,** No. 8, 75A–76A (August 1960); "The Literature of Chemical Health and Safety," 182nd National Meeting, American Chemical Society, New York City, August 27, 1981. See also Symposium on Information Sources for Specialized Knowledge in Chemical Health and Safety (20 papers), 186th ACS National Meeting, August 29, 1983–September 2, 1983, Washington, D.C.

9. National Safety Council, 444 North Michigan Avenue, Chicago, IL 60611.

10. American Industrial Hygiene Association, 475 Wolf Ledges Parkway, Akron, OH 44311.

11. Toxicology Information Response Center, Oak Ridge National Laboratory, P.O. Box X, Building 7509, Oak Ridge, TN 37830.

12. National Institute for Occupational Safety and Health, Cincinnati, OH 45226.

13. "Identification of the Fire Hazards of Materials," NFPA, No. 704-M. (Available from National Fire Protection Association, Batterymarch Square, Quincy, MA 02269.)

14. H. C. Hodge and J. H. Sterner, "Tabulation of Toxicity Classes," *Am. Ind. Hyg. Assoc. Q.* **10,** No. 4, 93 (December 1949). See also Gosselin, Hodge, Smith, Gleason, *Clinical Toxicology of Commercial Products, Acute Poisoning,* 4th ed., Williams & Wilkins Co., Baltimore, 1976 (inside cover).

15a. "Safety Solvent Fatality, Case History No. 442," *Case Histories of Accidents in the Chemical Industry,* Vol. 1, Chemical Manufacturer's Association, Washington, D.C., 1962.

  b. C. R. Fagen, "Two Overcome By Fumes in Chemical Spill SNAFU-MAD AV Closed By Toxic Cloud," *New York Post,* July 29, 1983, p. 13.

16. S. Winstein, "Communications to the Editor—Bicycloheptadiene Dibromides," *J. Am. Chem. Soc.* **83,** 1516–1517 (March 20, 1961).

17. "Vapors Fatal," (Letters to Editor), *Chem. Eng. News* **54,** No. 36, 5 (August 30, 1976).

18. "Glycol Ethers," Current Intelligence Bulletin 39, May 2, 1983. (Available from National Institute for Occupational Safety and Health, Cincinnati, OH 45226.)

19. L. M. Gibbs, *Love Canal, My Story,* State University of New York Press, Albany, N.Y., 1982.

20. R. Cook, *Fever,* Putnam, New York, 1982.

21. *EPA J. (Toxics),* **4,** No. 8 (September 1978). (Available from the U.S. Environmental Protection Agency, Washington, D.C. 20460.)

22. Y. Sa Woda, "Correlation of Pathological Findings With Computerized Tomographic Findings After Acute Carbon Monoxide Poisoning," *New Eng. J. Med.* **308,** No. 21 (May 26, 1983).

BIBLIOGRAPHY

"Annual Plan, Fiscal Year 1982," NTP-81-94, National Toxicology Program, March 1982. (Available from Public Information Office, National Toxicology Program, P.O. Box 12233, Research Triangle Park, NC 27709.)

Atherley, G. R. C., *Occupational Health and Safety Concepts,* Applied Science Publishers, London, 1978.

Bandal, S. K., Goldberg, L., Marco, G., and Leng, M., *The Pesticide Chemist and Modern Toxicology,* ACS Symposium Series 160, American Chemical Society, Washington, D.C., 1981.

Branson, D. R., and Dickson, K. L., "Aquatic Toxicity and Hazard Assessment," ASTM/STP737, September 1981; "Estimating the Hazard of Chemical Substances to Aquatic Life," ASTM TP657, American Society for Testing & Materials, Philadelphia, Pennsylvania, 1978.

Bunt, R., "The U.S. Fight Against Chemical War," *Wall Street Journal,* January 4, 1982, p. 31; "On the Agenda: Yellow Rain" Review and Outlook, *Wall Street Journal,* March 24, 1982, p. 26.

Caglioti, L., *The Two Faces of Chemistry,* MIT Press, Cambridge, Mass., 1983.

Choudhary, G., *Chemical Hazards in the Workplace,* ACS Symposium Series 149, American Chemical Society, Washington, D.C., 1981.

Clayton, G. D., and F. E. Clayton, *Patty's Industrial Hygiene and Toxicology,* 3rd ed. (in 5 volumes), Wiley–Interscience, 1978–1980.

Documentation of the Threshold Limit Values for Substances in Workroom Air, 4th ed., American Conference of Governmental Industrial Hygienists, (revised frequently). (Available from ACGIH, 6500 Glenway Avenue, Bldg. D-5, Cincinnati OH 45211.)

"Fiscal Year 1983 Annual Plan," NTP-82-119, January 1983, National Toxicology Program. (Available from the National Toxicology Program, P.O. Box 12233, Research Triangle Peak, NC 27709.)

Hazardous Waste Options, 16-mm sound color motion picture, 1981. (Available from Stuart Finley, 3428 Mansfield Road, Falls Church, Virginia.)

"Highest Priority," Review and Outlook, *Wall Street Journal,* January 7, 1982 p. 20.

Hopke, P. K., "Multitechnique Screening of Chicago Municipal Sewage Sludge for Mutagenic Activity," *Eng. Sci. Tech.* **16,** No. 6, 140–147 (March 1982).

Jones, C. J., "The Ranking of Hazardous Materials by Means of Hazard Indices," *J. Hazard. Mater.* **2,** No. 4, 363–389 (November 1978).

Large, A. J., "Of Mice and Men—Toxic Waste Fights Start in Laboratories, But How Good are the Testing Methods?" *Wall Street Journal,* June 21, 1983, p. 60.

Lowrence, W. W. (Ed.), *Assessment of Health Effects at Chemical Disposal Sites,* Kaufmann, Los Altos, Calif., 1981.

Mayer, C. E., "Report Paints Grim Picture of Asbestos," *Washington Post,* December 30, 1981, pp. D8–10.

McGrady, Pat, Sr., *The Persecuted Drug. The Story of DMSO,* Charter Books, New York, 1979.

Miller, J. A., "Are Rats Relevant?" *Sci. News* **112,** 12–13 (July 2, 1977).

Moeschlin, S., "Outstanding Symptoms of Poisoning," *Diagnosis and Treatment,* Grune & Stratton, New York, 1965, pp. 644–678.

National Research Council, *Identifying and Estimating the Genetic Impact of Chemical Mutigens,* National Academy Press, Washington, D.C., 1983.

Office of Technology Assessment, *The Role of Genetic Testing in the Prevention of Occupational Disease, Summary,* OTA-BA-195, Congress of the United States, Washington, D.C. 20510. April 1983.

Plimmer, J. R., *Pesticide Residues and Exposure,* ACS Symposium Series 182, American Chemical Society, Washington, D.C., 1982.

"Registry of Toxic Effects of Chemical Substances," prepared for NIOSH. (Revised annually); also available on microcards (quarterly), National Institute for Occupational Safety and Health, Cincinnati, Ohio.

"Review of Current DHHS, DOE, and EPA Research Related to Toxicology, Fiscal Year 1982," NTP-82-040, National Toxicology Program, June 1982. (Available from the National Toxicology Program, DHHS, P.O. Box 12233, Research Triangle Park, NC 27709.

Ross, S. S., *Toxic Substances Sourcebook* (Environment.) (Available from Information Center, 292 Madison Avenue, New York, NY 10017.)

Scanlan, R. A., and Tannenbaum, S. R., *N-Nitroso Compounds,* ACS Symposium Series 174, American Chemical Society, Washington, D.C., 1981.

Sittig, M. (Ed.), *Priority Toxic Pollutants, Health Impacts, and Allowable Limits,* Noyes Data Co., Park Ridge, N.J., 1980.

Spiro, T. G., and Stigliani, W. M., *Environmental Issues in Chemical Perspective,* State University of New York Press, Albany, N.Y., 1980.

Stahl, W. H., *Compilation of Odor and Taste Threshold Value Data, E-18 Committee on Sensory Evaluation of Materials and Products,* American Society for Testing and Materials, Philadelphia, 1973.

Tabor M., "Minimizing the Menace of OTC Drugs," *Occup. Health Safety,* **51,** No. 5, 14–19 (May 1982).

"WHO Consultation on the Implementation of WHA Resolution 30.47: Evaluation of the Effects of Chemicals on Health," Geneva, Switzerland, May 1, 1978–May 5, 1978. (Available from World Health Organization, 1211 Geneva, 27, Switzerland.)

Yoder, T. A., "Chemicals That May Affect Our Bodies, Babies, and Biology," *Chemical Newsletter,* National Safety Council, Chicago, August 1978, pp. 1–2.

TOXICITY CONSIDERATIONS

"An Assessment of the Health Risks of Morpholine and Diethyl-Aminoethanol," 40 pp., 1983. (Available from the Committee on Toxicology, National Research Council, Washington, D.C. 20418. The committee studied toxicity of these materials in boilers, and concluded that morpholine and DEAE should not be used simultaneously in boiler systems, due to possible nitrosamine formation. The report is available from the committee.)

Annotated bibliography of "Effects of Environmental Chemicals on the Immune System," on "Chemical Waste Disposal," "Health Aspects of Urea/Formaldehyde Compounds," "Toxicity of Vanadium Compounds" and "Health Aspects of Asbestos in Air," 1982. (Available from the Federation of American Societies for Experimental Biology, 9650 Rockville Pike, Bethesda, MD, 20814.)

Anon., "The Effects of Toxic Substances and Physical Agents on the Reproductive System," (last page, unnumbered), *The Chemical Worker* (*monthly*), International Chemical Workers Union. No. 8 (August 1982).

"Asbestos-Containing Materials in School Buildings: A Guidance Document, Part 1, 000090," March 1979, Office of Toxic Substances, U.S. Environmental Protection Agency, Washington, D.C.

Brown, M. H., *Laying Waste,* Pantheon Books, New York, 1980.

Bulugahapitya, D. T. D., "Near-Fatal Asphyxia By a Toy Shopping Bag," *Brit. Med. Jour.* **285,** 203, (July 24, 1982).

Chemical Hazards in the Workplace: Measurement and Control, G. Choudhary, (Ed.), ACS Symposium Series 149, American Chemical Society, Washington, D.C. 1981.

"Chemicals and Cancer," in *Cancer, the Misguided Cell,* D. M. Prescott and A. S. Flexer (Eds.), Scribner's, New York, 1982, Chap. 7, pp. 137–173 (written for popular audience).

CIS Abstracts (in English) of reference texts in various languages providing information about toxic or hazardous chemicals. Information Search, Retrieval Service of the International Occupational Safety and Health Information Centre, Bureau of International du Travail, CH-1211, Geneve 22, Switzerland, 1982.

Current Approaches to Occupational Health, A. W. Gardner, (Ed.), John Wright—PSG Inc., Littleton, Mass., 1982.

Dawson, G. W., "Risk Management and the Landfill in Hazardous Waste Disposal," *J. Hazard. Mater.* **8,** No. 1, 69–84 (June 1983).

Diet, Nutrition and Cancer, Committee on Diet, Nutrition and Cancer, National Research Council, National Academy Press, Washington, D.C., 1982.

"Draft, Regulatory Impact Analysis and Regulatory Flexibility Analysis of the Hazard Communication Proposal," PB 82-194549, OSHA.

"Economic Implications of Regulating Chlorofluorocarbon Emissions from Nonaerosol Applications," Rand Corp., PB 82-15623, NTIS.

"Fiscal Year 1982, Review of Current DHHS, DOE, and EPA Research Related to Toxicology," *National Toxicology Program,* NTP-82-040, June 1982. (Available from the National Toxicology Program, P.O. Box 12233, Research Triangle Park, NC 27709.)

Freundt, K. J., "Mixed Exposures to Chemical Hazards," *Occup. Health Safety,* 10–13, 39–42 (August 1982).

Gillies, M. T., *Drinking Water Detoxification,* Noyes Data Corp., Park Ridge, N.J., 1978.

Haque, R., *Dynamics, Exposure and Hazard Assessment of Toxic Chemicals,* Ann Arbor Science Publishers, Inc. Ann Arbor, 1980.

Hazardous Chemicals Data Book, G. Weiss, (Ed., Environmental Health Review No. 4, Noyes Data Corp., Park Ridge, N.J., 1980 (reviews 1350 chemicals).

Hazardous Waste Options, 16-mm sound color motion picture, 1982, 28 min. (Available from Stuart Finley, Inc., Falls Church, VA 22041.)

Health Impacts of Polynuclear Aromatic Hydrocarbons, A. W. Pucknat, (Ed.), Environmental Health Review No. 5, Noyes Data Corp., Park Ridge, N.J., 1981.

Jennings, A. A., "Profiling Hazardous Waste Generation for Management Planning," *J. Hazard. Mater.* **8,** No. 1, 69–84 (June 1983).

Jones, C. J., and McGugan, P. J., "Evaporation of Mercury from Domestic Waste Leachate," *J. Hazard. Mater.* **2,** No. 3, 253–258 (August 1978).

Management of Assessed Risk for Carcinogens, W. J. Nicholson, (Ed.), New York Academy of Sciences, New York, 1981.

Mathias, C. G. T., "Contact Dermatitis: When Cleaner is not Better," *Occup. Health Safety,* pp. 45–50, (January 1984) (5 refs.).

Mehlman, M. A., *Hazards from Toxic Chemicals: Proceedings of the Second Annual Conference on the Status of Predictive Tools in Application to Safety Evaluation,* Pathotox Publishers, Inc., Park Forest South, Ill., 1978.

National Research Council, *Kepone/Mirex/Hexachlorocyclopentadiene: An Environmental Assessment,* National Academy of Sciences Press, Washington, D.C., 1978.

National Research Council, *Toxicity Testing: Strategies to Determine Needs and Priorities,* Study conducted for National Toxicology Program, National Academy Press, Washington, D.C. 20418, March, 1984.

"National Toxicology Program Chemical Registry Handbook," PB82-142977, NTP, Research Triangle Park, NC 27709.

N-Nitroso Compounds, R. A. Scanlan and S. R. Tannenbaum (Eds.), ACS Symposium Series 174, American Chemical Society, Washington, D.C., 1981.

Pastor, A., "Administration Takes Steps to Ease Controls on Cancer-Causing Agents," *Wall Street Journal,* December 21, 1982, p. 7.

Pesticide Manufacturing and Toxic Materials Control Encyclopedia, M. Sittig, (Ed.), Chemical Technology Review No. 168, Noyes Data Corp., Park Ridge, N.J., 1980.

Plans for Clinical and Epidemiologic Follow-Up After Area-Wide Chemical Contamination, Proceedings of an International Workshop, Washington, D.C. March 17, 1980–March 19, 1980, National Academy Press, Washington, D.C., 1982.

Plimmer, J. R., *Pesticide Residues and Exposure,* ACS Symposium Series 182, American Chemical Society, Washington, D.C. 1982.

Plunkett, E. R., *Handbook of Industrial Toxicology,* Chemical Publishing Co., New York, 1976.

Polakoff, P. L., "Occupational Cancer: Avoidable Risk," *Occup. Health Safety,* 23–26 (August 1982).

Purves, D., *Trace Element Contamination of the Environment,* Elsevier, New York, 1977.

Pye, A. M., "A Review of Asbestos Substitute Materials in Industrial Applications, *J. Hazard. Mater.* **3,** No. 2, 125–148 (September 1979).

Robinson, R., and Stott, R., *Medical Emergencies, Diagnosis and Management,* 3rd ed., William Heinemann Medical Books, London, 1980.

Sanini, B. S., and Williams, A. M., "Occupational Exposure to Asbestos Fibers Resulting from Use of Asbestos Gloves," *Amer. Ind. Hyg. Assoc. J.,* **42,** No. 12, 872–875 (December 1981).

Schuetzle, D., *Monitoring Toxic Substances,* ACS Symposium Series 94, American Chemical Society, Washington, D.C., 1979.

Sittig, M. *Hazardous and Toxic Effects of Industrial Chemicals,* Noyes Data Corp., Park Ridge, N.J., 1979.

Sittig, M., *Toxic Metals, Pollution Control and Waste Production,* Noyes Data Corp., Park Ridge, N.J., 1976.

Smith, R., "Alcohol and Alcoholism: The Politics of Alcohol," *Brit. Med. Jour.* **284,** 1392–1394 (May 8, 1982).

Tabor, M., "Going to the Problem's Source: Why Workers Drink on the Job, and Why They Shouldn't: Alcohol, Prescription Drugs, Stress and Chemical Exposures Form a Serious Threat to Worker Safety," *Occup. Health Safety,* 30–35 (August 1982).

"The Alkyl Benzenes," PB82-160334, National Research Council.

The Pesticide Chemist and Modern Toxicology, S. K. Bandal, G. J. Marco, L. Golberg, and M. Leng ACS Symposium Series 160, American Chemical Society, Washington, D.C. 1981.

Touhill, C. J., "Hazardous Waste Management at Abandoned Dump Sites—Evolving Perspectives," *J. Hazard. Mater.* **6,** No. 3, 261–266 (May 1982).

Tu, A. T., *Survey of Contemporary Toxicology,* Wiley, New York, 1981.

Tyrer, F. H., and Lee, K., *A Synopsis of Occupational Medicine,* John Wright—PSG Inc., Littleton, Mass., 1979.

"VOC (Volatile Synthetic Organic Chemicals) in Drinking Water," *Fed. Reg.* **47,** 9350 (March 4, 1982); *Fed. Reg.* **47,** 24,756 (June 8, 1982).

West, A. S., "Risk-Benefit Trade-Offs Under the Toxic Substances Control Act," *Chem. Eng. Prog.,* 88–92 (August 1982).

Willman, J. C., "Case History: PCB Transformer Spill, Seattle, Wash.," *J. Hazard. Mater.* **1,** No. 4, 361–372 (March 1977).

Wimer, W. W., Russell, J. A., and Kaplan, H. L., *Alcohols Toxicology,* Noyes Data Corp., Park Ridge, N. J., 1983.

Toxicity

"ACS Statement on Volatile Synthetic Organic Chemicals in Drinking Water," Department of Public Affairs, American Chemical Society Washington, D.C. (Submitted to the EPA on August 20, 1982.)

Bandal, S. K., *The Pesticide Chemist and Modern Toxicology,* ACS Symposium Series 160, American Chemical Society, Washington, D.C., 1981.

Barrons, K. C., "Toxicity vs Hazard: Dioxin and the Oleander," *Wall Street Journal,* June 17, 1983, p. 26.

"EPA Implementation of Selected Aspects of the Toxic Substances Control Act," General Accounting Office, Washington, D.C., PB83-156125, NTIS, 1983.

Glickman, N. W., et al., *Accidental Poisoning of Children by Veterinary Drugs,* pp. 1–2; CHEMTREC 800 Watts Line Service [on (800)424-9300] p. 3; updated toxic emergency phone number listing available, p. 4; citations from the recent literature, pp. 4–13; *Bulletin of National Clearinghouse for Poison Control Centers* **26,** No. 2 (April 1982–June 1982). (This publication is terminating publication after this issue due to budget constraints.)

Hamilton, A., Hardy, H., and Finkel, A. J., *Industrial Toxicology,* 4th ed., John Wright—PSG Inc., Littleton, Mass., 1982.

Harris, A., "Checks Came in the Mail, But the Poison is Still in the Catfish," *Washington Post,* June 4, 1983, p. A2. (Triana, Alabama residents participated in out-of-court settlement of $24 million from dumping at least 475 tons of DDT into Indian Creek over 17 years. Costs of DDT clean-up could run as high as $90 million.)

Helliker, K. P., "Some Experts Say Gulf States are Ignoring Serious Problems in Toxic-Waste Disposal," *Wall Street Journal,* June 17, 1983, p. 26.

"Increasing the Usefulness of Acute Toxicity Tests, Environmental Research Laboratory," Duluth, Minnesota, PB83-151712, NTIS.

"Investigation of Agents which Are Newly Suspected as Occupational Health Hazards: Epichlorohydrin," NIOSH, PB83-153874, NTIS.

"National Pesticide Information Retrieval System (NPIRS): NPIRS—Facts," Purdue University, PB83-148072, NTIS, 1983.

"Occupational Health Guidelines for Chemical Hazards," Arthur D. Little, Inc., Cambridge, Mass., PB83-154609, NTIS.

Pesticide Residues and Exposures, ACS Symposium Series 182, American Chemical Society, Washington, D.C., 1982.

"Registration and Procedures for Safety Reassessment for Drugs and Chemicals Used in Agriculture: Some Economic Considerations," Economic Research Service, Washington, D.C., PB83-149716, NTIS, 1983.

"Scientific Papers at Symposium on Chemical and Other Environmental Toxicity Factors Affecting the Eye, Ear, and Other Special Senses," Environmental Health Perspective Series, (NIH 82-218), U.S. GPO, Washington, D.C.

Seamer, J. H., and Wood, M., *Safety in the Animal House,* Laboratory Animal Handbooks, 2nd rev. ed., Laboratory Animals, Inc., 33 Furrowfelde, Kingswood, Basildon, Essex SS16 5HA, England.

Sittig, M., *Handbook of Toxic and Hazardous Chemicals,* Noyes Data Corp., Park Ridge, N.J., 1981.

"Toward a Federal/State Partnership in Hazardous Materials Transportation Safety," Department of Transportation, Washington, D.C., PB83-121558, NTIS, 1983.

"Toxic Waste Disposal Companies Facing A Profitable Future Despite Bad Publicity," *Wall Street Journal,* June 10, 1983, p. 33.

"Toxic Waste Dumper Jailed," *Washington Post,* July 23, 1983, p. A3. (President of water service franchise sentenced to 90 days in jail for illegally disposing of toxic waste into storm sewers in Log Angeles, California.) "Questions about the Environmental Protection Agency," Round the World, Correspondent, *The Lancet,* March 20, 1982, p. 672; see also "The Carcinogenicity of Formaldehyde," *The Lancet,* 1981, p. 576.

OCCUPATIONAL HEALTH CLINICS

Since most medical schools in the United States, even today, do not require their students or graduates to become deeply involved with occupational health problems, the following list of clinics may be useful in locating physicians and others with specific knowledge in any chemical exposure, including hazardous-waste-site exposures:

California

Barlow—U.S.C. Occupational Health
    Center
200 Stadium Way
Los Angeles, CA 90026

Occupational Health Clinic
San Francisco General Hospital
1001 Potrero Avenue
San Francisco, CA 94110

Connecticut

Yale Occupational Medicine Program
333 Cedar Street
New Haven, CT 06510

Illinois

Occupational Medicine Clinic
Cook County Hospital
720 South Wolcott
Chicago, IL 60612

Kentucky

University of Kentucky Medical
    Center—Pulmonary Division
800 Rose Street
Lexington, KY 40536

Maryland

Center for Occupational and
    Environmental Health
Johns Hopkins University
3100 Wyman Park Drive, Bldg. 6
Baltimore, MD 21211

Occupational Medicine Clinic
Baltimore City Hospital
4940 Eastern Avenue
Baltimore, MD 21224

Massachusetts

Occupational Health and
    Environmental Health Center
Bringham and Women's Hospital
721 Huntington Avenue
Boston, MA 02115

Occupational Health Clinic
Norfolk County Hospital
2001 Washington Street
South Braintree, MA 02184

Occupational Health Service
Department of Family and Community
    Medicine
University of Massachusetts Medical
    Center
55 Lake Avenue North
Worcester, MA 01605

Occupational Medicine Clinic
Cambridge Hospital
1493 Cambridge Street
Cambridge, MA 02139

Occupational Medicine Clinic
Massachusetts General Hospital
Fruite Street
Boston, MA 02146

Michigan

University of Michigan
Occupational Health Clinic
School of Public Health II
Room M-6012
Ann Arbor, MI 48106

New Jersey

Occupational/Environmental Disease
    Clinic
New Jersey Department of Health
CN-360
Trenton, NJ 08625

Occupational Medicine Group
714 Broadway
Paterson, NJ 07514

New York

Mt. Sinai Medical Center
Occupational Medicine Clinic
100th Street & Fifth Avenue
New York, NY 10029

Occupational Health Clinic
Montefiore Hospital
111 East 210th Street
Bronx, NY 10467

Texas

University of Texas Medical Center
School of Public Health
Houston, TX 77025

ADDITIONAL INFORMATION RESOURCES

American Chemical Society
Safety and Health Referral Service
1155 16th St., N.W.
Washington, D.C. 20036
(202)872-4511 (B. Gallagher)

American Society of Law & Medicine
765 Commonwealth Avenue
Boston, MA 02215

Asbestos Victims of America
P.O. Box 559
Capitaola, CA 95010
(408)476-3646

Asbestos Victims of America
312 Essex Street
Gloucester City, NJ 08030
(609)456-5695

Carcinogen Informational Project Program
P.O. Box 6057
St. Louis, MO 63139

International Society for Respiratory Protection
Box 7567
St. Paul, MN 55119

Society of Environmental Toxicology
and Chemistry
P.O. Box 352
Rockville, MD 20850

4
Fires and Explosions

The tendency in many instances to associate chemicals with toxicity (as discussed in Chaps. 2 and 3) frequently overlooks the fundamental fact that most chemical substances will burn, decompose, or explode if improperly handled or disposed of. This is especially true of the liquids, which constitute a major volume of the waste in many industries, since liquids, more than solids or sludges, tend to eventually escape from their containment and find their way into unwanted directions, whether it be a chemical reaction, a water supply, a stream, or other difficult-to-control location. The author vividly recalls being asked to judge the safety of a small house, a residence for two elderly persons, about 1000 ft (305 m) from a large above-ground gasoline storage tank which had been overfilled from a barge, and where gasoline (perhaps 300 gal) (1130 liters) had soaked into the ground. On entering the house, the odor of gasoline was very strong; use of a vapor tester indicated between 750 and 900 ppm in the basement furnace room where a coal fire was used as a heat source. On our recommendation, the fire was extinguished in the furnace and the elderly couple sent to Florida for 2 months until the spring thaw would permit reevaluation of the problem and probably reestablishment of residency. We cite the above, which occurred over 20 years ago, to indicate that liquids will travel significant distances, even in the winter conditions of New York State.

In the handling and disposal of flammable and combustible materials, certain fundamentals must be recognized and observed if difficulty is to be avoided. Although fire is probably mankind's first chemistry, even today our knowledge, and the application of that knowledge, is too limited when we read of a jet aircraft forced to make an emergency landing due to a fire in a washroom (exact cause not known but believed either electrical malfunction or careless disposal of smoking materials), resulting in the loss of 20 lives, most believed due to smoke inhalation from the burning or smoldering seats and other plastic components of the cabin; or the recent closing of Madison Avenue in Manhattan (July 28, 1983) when 55 gal of a chlorinated solvent reacted with an aluminum hand-pump, causing evacuation of a 12-story office building and fear of more extensive damage to the building.

In the United States, the standards on flammable liquids set up by the National Fire Protection Association, known as NFPA 30, identify materials and provide the test methods for the proof of the identity insofar as fire characteristics are involved. Why the RCRA and SUPERFUND must introduce a relatively new term "ignitability" to wastes, to the already confusing "flammable," "inflammable," and "combustible" designations, we can only speculate in terms of the legal approach.

For disposal purposes (as noted in Chap. 7), "ignitability" is defined by the RCRA as follows:

1. For liquids, as a material having a flash point less than 60°C (140°F) by the specified methods.
2. For nonliquids, as a material capable of ignition under normal conditions of spontaneous and sustained combustion.
3. For an ignitable compressed gas, per Department of Transportation regulations, and/or
4. For an oxidizer per Department of Transportation regulations. The EPA hazard code for ignitability is "I", with EPA hazardous-waste number D001.

In addition to ignitability, another very important consideration from a fire and explosion viewpoint is "reactivity." The EPA defines this characteristic in terms of:

Normally unstable—reacts violently.

Reacts violently with water.

Forms explosive mixture with water.

When mixed with water, generates toxic gases, vapors, or fumes.

Contains cyanide or sulfide and generates toxic gases, vapors, or fumes at pH between 2 and 12.5.

Capable of detonation if heated under confinement or subjected to strong initiating source.

Capable of detonation at standard temperature and pressure.

Listed by Department of Transportation as Class A or Class B explosive.

Reactive materials have been assigned the EPA hazardous-waste number D003.

Regardless of the terms used, the fundamentals of ignition cannot be overlooked by prudent and reasonable persons in connection with chemically related wastes. It is difficult to understand why our society assigns such a *low* importance to fire *prevention* and *protection,* in contrast to fire *fighting,* until actual flame or smoke is observed.

At the risk of seeming overly elementary, a simple review of fire, in the context of chemical substances, appears in order. Combustion is the violent combination of oxygen or some other oxidizing material with a combustible substance (i.e., combustion will furnish the oxidation potential and heat required for burning the substance). As mentioned in Chapter 1, we believe that, for a fire to be sustained, the four elements of the fire "square" are necessary—namely, fuel available in the appropriate condition, such as gas or vapors; oxygen (usually air, but other oxidizers such as the halogens and nitrogen oxide are in order); an ignition source or spark of sufficient energy potential to initiate the reaction; and, for a sustained fire, a chain reaction in which sufficient heat is transferred as part of the system to maintain the other elements of the "square." Fire, of course, is exothermic, and even materials with low volatility and little tendency to burn, will do so if heated or atomized.

Since many hazardous wastes are liquids, classification of the various classes of liquids, as prescribed by the National Fire Protection Association NFPA 30 in 1981, should be well understood:

Flammable Liquids

Class 1A is defined as having a flash point below 73°F (22.8°C) and a boiling point below 100°F (37.8°C).

Class 1B is defined as having a flash point below 73°F (22.8°C) and a boiling point at or above 100°F (37.8°C).

Class 1C is defined as having a flash point at or above 73°F (22.8°C) and below 100°F (37.8°C).

Combustible Liquids

Class II is defined as having a flash point at or above 100°F (37.8°C) and below 140°F (60°C).

Class IIIA is defined as having a flash point at or above 140°F (60°C) and below 200°F (93.4°C).

Class IIIB is defined as having a flash point at or above 200°F (93.4°C).

Flash point, that is, the temperature at which the vapors directly over the surface of a liquid will flash or ignite, can be determined by several procedures and forms of apparatus, as specified by the various codes of the American Society of Testing and Materials. Fire point, sometimes referred to, is usually a few degrees higher than the flash point, and is the temperature at which the flash fire will continue to burn. (For a complete understanding of these tests, the reader is referred to the publications of the National Fire Protection Association, Batterymarch Square, Quincy, MA 02269.)

Air, which is usually the oxidation source in most fires, supports combustion readily but at a controlled rate. When the oxygen percentage in air is increased even a few percents, the rate of burning increases radically, and very great danger is encountered (see Fig. 1.4). Other oxidizing agents include chlorine, fluorine, ozone, nitrogen oxides, the interhalogens, such as ClF_3, and other substances containing oxygen, such as $KMnO_4$, $K_2Cr_2O_7$, the chlorates, and perchlorates. All may be in waste sites.

The third aspect of liquid or gas flammability, namely the necessity for the liquid or gas to either be in a mist or within the "flammable" range (sometimes called "explosive" range), is the salvation in many situations of waste sites, since unless

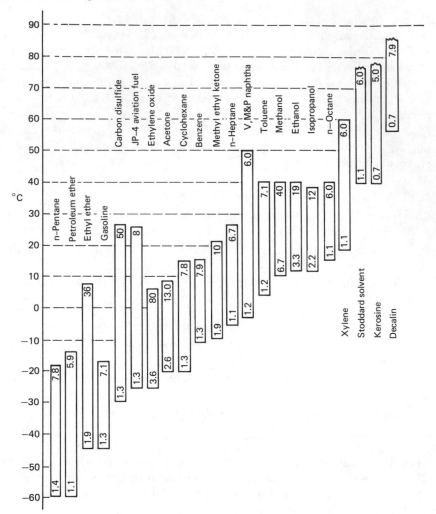

Figure 4.1. Relation of temperature with flammable range of common chemicals.

the lower flammable limit is present, ignition and burning is not possible. For materials which are highly volatile, such as the ethers, acetone, benzene, naphthas, and dioxine, the lower flammable limit (known as LFL) is present at normal conditions over the surface. The application of the ignition source of sufficient energy, plus the ambient air, are then meeting all requirements for a fire, which, once the chain reaction occurs, is self-sustaining. If, on the other hand, the substance is "too rich" to burn, by being above the "upper flammable limit" or UFL, fire is not possible, until it is diluted down to within the flammable range. For some substances, such as diborane and ethylene oxide, this may be a very wide range; for other materials, such as toluene and motor-fuel gasoline, the range may be only a few percent, Figures 1.4, 4.1, and 4.2 give typical values for flammable range.

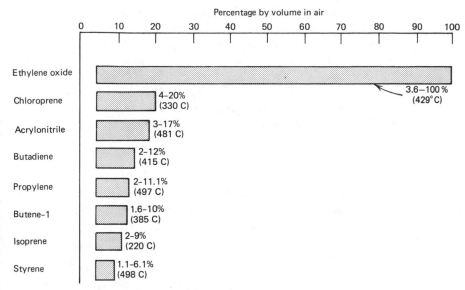

Figure 4.2. Flammable ranges and autoignition temperatures of certain common monomers.

[For additional values, consult the publications of the NFPA or *Safety and Accident Prevention in Chemical Operations,* 2nd ed., H. H. Fawcett and W. S. Wood, (Eds.), Wiley–Interscience, New York, 1982, Appendix 1, pp. 865–875, or your insurance carrier or local or state fire marshal's office.]

If there is any hint or suspicion that a flammable gas or vapor is present, a survey with a portable flammable gas detector is in order. It would be wise to follow the recommendations of the National Safety Council, Chemical Section, "Safety and Health Information Sheet," CHM 83-1, August 1983, which is summarized below.

The information sheet discusses the two basic types of instruments—aspirated or nonaspirated (diffusion). Aspiration can be manual, using a squeeze bulb, or can be by electric pump. Some nonaspirated types have the capability for aspiration as an added feature. The detector is usually a catalytic element which is part of an electrical circuit. The gas to be tested is brought to the detector by aspiration or diffusion depending upon the type of instrument. If the gas is flammable, it will react with the catalytic element producing a change in the electrical resistance, which will then produce a deflection in the instrument meter. The meter reads in percentages of the lower flammable limit (or LFL). The instrument is calibrated with a known flammable gas. It should be recalibrated on a schedule, such as once a month, to detect changes in the instrument which could give inaccurate results.

To use the instrument, the following steps are recommended:

1. Read and understand the instrument manufacturer's instructions.

2. Be certain the instrument has been recently recalibrated or within the scheduled time for recalibration.

3. "Zero" the instrument in a clean (uncontaminated) air.

4. Size-up the sampling job, by asking such questions as: What flammables are likely to be present? How can a representative sample of the atmosphere be obtained? It must be appreciated that most vapors are heavier than air, and tend to "pocket" or seek lower areas. This is especially true when investigating tanks and other enclosed spaces. [See T. A. Kletz, "Hazards in Chemical Systems Maintenance: Permits," and also W. S. Wood, "Safe Handling of Flammable and Combustible Materials," in *Safety and Accident Prevention in Chemical Operations,* 2nd ed. H. H. Fawcett and W. S. Wood, (Eds.), Wiley–Interscience, New York, 1982, pp. 807–836 and pp. 339–356.]

5. If ventilation is in use, sample both with it off and on.

6. Watch the instrument meter at all times during the sampling, since a rich mixture may cause a momentary positive reading which may then drop back to zero as the oxygen inside the meter is consumed. Unless the instrument operator saw this momentary meter deflection, he or she might assume incorrectly that the atmosphere is safe.

Some pitfalls to avoid, include "real-world" aspects:

1. Failure to recalibrate an instrument on a scheduled basis with records to back-up the recalibration.

2. Failure to obtain a representative sample of the atmosphere.

3. Sampling an atmosphere that would damage the instrument or detector, such as atmospheres containing acid, lead, or the silicones.

4. Permitting liquid or condensate to enter the instrument or contact the detector. This can result in inaccurate readings or seriously damage the detector. Sampling close to a liquid surface, sampling an atmosphere containing steam, or sampling a hot atmosphere containing condensable vapors is to be avoided.

The following must be considered if results are to be properly interpreted: Since the instrument is calibrated on a known flammable gas, other flammable gases and vapors with different molecular weights can often give significantly different readings. Consequently, unless the flammable gas in the atmosphere being sampled is the same as the calibration gas, the meter reading may not be correct. The meter reading can also be very sensitive to how the instrument is operated. Consequently, meter readings should only be regarded as estimated, and should be repeated several times to ensure better and more accurate readings. For additional precautions where "hot work" or confined space entry work is involved, the recommendations of persons skilled in the use of such instruments should be taken very seriously. In general, if readings above 10% of the LFL are obtained, hot work or tank or vessel entry would be very unwise, if not dangerous.

It should be noted that some detectors for flammable materials also may be recalibrated for use as industrial hygiene survey instruments, with a different range on the meter. Most reputable instrument manufacturers can recommend the particular instrument most suitable to the applications which may be anticipated.

In addition to the "hot wire" type instrument described above, thermal conductivity, diffusion rate, specific gravity, and infrared spectrum instruments are coming into use, and may find application in waste-site surveys and investigations.

The growing use of plastic containers for gasoline and other flammable liquids, some with the approval of Factory Mutuals Insurance ratings, brings into the situation an additional hazard not previously present. When "safety cans," so-called since their all-metal construction, together with self-closing closure and a "flame arrestor" to allow flashes into or out of the container, were considered standard, there was much greater safety from a spill or ignition occurring (see Figure 4.3). There is a tendency to use plastic containers for waste in many laboratories and factories. When these containers are placed in a "secure landfill," they are more subject to puncture and to fire than the metal containers. If a fire occurs, in addition to the toxic components in the smoke from the solvent fire, the decomposition or pyrolysis products of plastics are added to the problem. Table 4.1, compiled by R.H.L. Howe in 1972, notes the difference between products of combustion at 350°F (177°C) and 1600°F (871°C). (For more information on this subject, the reader is referred to the 10-volume series *Fire Safety Aspects of Polymeric Materials*, a report by the National Materials Advisory Board of the National Academy of Sciences published in 1980 by Technomic Publishing Company, Westport, CT 06880, or to the National Bureau of Standards.) Fire in discarded automobile tires piled over a five acre area near Winchester, Virginia in October 1983 resulted in severe air pollution to parts of three states, and control measures that have cost over $500,000 for the first two months. A petroleum residue is being collected and recycled to prevent pollution of the Potomac River.

Fire Toxicity

Although we have discussed toxicity (Chaps. 2 and 3) and the necessity for respiratory protection (Chap. 6) and other personal protective equipment (Chap. 5) in

Flame
Arrester

Figure 4.3. Metal "safety can" showing flame arrestor screen and self-closing top. Such cans have an excellent record in storage and handling of flammable liquids.

TABLE 4.1. Burning of Solid Plastic Materials

Type of Plastic	Combustion Products at 350°F (177°C)	Combustion Products at 1600°F (871°C)	Remarks
1. Phenolic	Aldehyde, phenol, CO, benzoates, phenol acetic acid, etc.	CO and CO_2	May have smoke at low-temperature ranges.
2. Amino (trimethylol melamine)	Aldehyde, NH_3, formic acid, CO, CO_2, cyanate, amide, etc.	CO, CO_2, NO_2, N_2, etc.	May be toxic at low temperature.
3. Polyester and alkyds	Phthalic anhydride esters, carboxylic acids, phthalic alkyd, CO_2, glycerides, etc.	CO and CO_2	May generate smoke at lower temperatures.
4. Polyvinyl acetate	Vinyl acetate, acetic acid, CO, and CO_2	CO_2, CO, and H_2O	May generate smoke at lower temperatures.
5. Polyvinyl chloride[a]	HCl, $COCl_2$, CO, benzene, etc.	CO_2, CO, HCl, nascent Cl_2, and $COCl_2$	Irritating, corrosive, and toxic.
6. Polystyrene	Styrene, partially burned carbon, benzene, ethylene, smoke, etc.	CO and CO_2	May generate smoke at lower temperatures.
7. Polyurethane	Tolylene diisocyanate, toluene, cyanate phenylhydantoic acid, etc., and smoke	Phenyl compounds, cyanate, CO_2, NO_2, and N_2	Toxic at high concentration at low-temperature burning.
8. Polytetrafluoroethylene	CF_4, CO_2, CO, and fumes	F_2 (at 750°F), CO_2, and COF_2	Toxic at high concentration.
9. Polyethylene	$CH_2{=}CH_2$, CO, CO_2, and some smoke	CO_2 and H_2O	May generate smoke at low temperatures.
10. Epoxy resins	No reaction or melted at most	CO, CO_2, and H_2O	May need temperature of 1600°F for complete combustion.
11. Polypropylene	Melted or no reaction	CO_2 and H_2O	May need temperature of 1600°F for complete combustion.
12. Petroleum coal tar resins: (1) straight and closed chain derivatives (asphalt) and (2) cyclic derivatives (tar).	(1) Benzene, phenols, benzoic acid, and smoke. (2) Naphthalenelike gases, benzene, HCN, carbazole, etc., and smoke.	(1) Benzoic acids, CO, CO_2, and H_2O. (2) CO, CO_2, and aromatic fumes.	(1) Requires high temperature, 2500°F, to burn. (2) Requires higher temperatures for complete combustion.
13. Urethane (ethyl carbonate) and polyurethane. Tolylene diisocyanate (see also 7).	CN, CO, CO_2, NO, NH_3, aldehyde, amide (HF if fluorocarbon was also used), and CNO	CO, CO_2, NO_2, N_2, etc., and HF	Other product depending on the temperature. If fluorocarbon is used as a blowing agent, HF will be generated.
14. Neoprene (2-chloro-1,3-butadiene)	CO, CO_2, HCl, unburnt particles, and smoke	CO_2, HCl, and fumes	Higher temperature than 2500°F is required for complete combustion.

TABLE 4.1.   *(Continued)*

Type of Plastic	Combustion Products at 350°F (177°C)	Combustion Products at 1600°F (871°C)	Remarks
15. Bakelite phenol or urea formaldehyde	CO, CO_2, CN, CNO, NH_3, aldehyde, amide, H_2O (phenol) benzoate, phenol acid, etc.	CO, CO_2, NO_2, N_2, H_2O, etc.	Also known as dimethylol ethylene urea or dimethylol ethyltriazone. Phenol formaldehyde is similar to phenolic resin.
16. Lucite (methyl methacrylate)	CO, CO_2, H_2O, acetic acid, etc.	CO_2 and H_2O	Low-temperature burning may generate methyl formate.
17. Plexiglass (acrylate and methacrylate)	CO, CO_2, H_2O, acetic acid, etc. (CN, CNO, NH_2, etc. if a nitrile is involved)	CO_2 and H_2O (NO_2 and N_2 if a nitrile is involved)	16 and 17 are quite similar. Sometimes a nitrile is involved.

^{a}For a discussion of PVC, see pages 79–81.

any detailed investigation of hazardous-waste sites, the importance of the subject of toxicity, especially if ignition occurs in the waste site, or during the removal of waste chemicals, is of great significance. The fire services and other emergency personnel are especially vulnerable to the toxic aspects of waste chemicals, when called to control an emergency. The wide variety of substances which may be encountered by persons having only limited understanding or expertise compounds the potential for serious injury, even in well-trained and well-equipped departments.

A special issue of *the International Fire Chief* magazine, Volume 29, Number 7, July 1983 is dedicated to toxicity as seen from the viewpoints of fire personnel. In introducing the series of articles, the editor of the magazine notes that toxicity still is an emerging issue, and that opinions of many on this subject are in a state of flux. This is a refreshing change for the better from the days when the traditional concept was that fire fighters were "smoke eaters," and that only weaklings wore breathing apparatus or considered the subject of toxicity seriously.

It was noted in the magazine issue that toxicity is only one of a handful of flammability characteristics that must be considered before materials are regulated. The limitations of smoke-toxicity experiments and burning-rate testing are real, and often can be properly challenged. The editor pleads for a constructive role among commercial interests in finding solutions openly and honestly.

In the article from the magazine, "Toxicity of Combustion Products," pages 12–15, Professor I. N. Einhorn notes that fire is a major societal problem in the United States where annually approximately 8000 persons die from fire-related exposures, and more than 20,000 serious injuries are attritutable to fire. Also, 50,000 additional persons exposed to fire require medical care, often for prolonged periods. The majority of the fire-related injuries and deaths result from exposure to combustion products, rather than from the fire itself. A complex mixture of gases, aerosols, particulates, and material fragments include carbon monoxide and dioxide, oxides of nitrogen, hydrogen cyanide, acid gases (such as HCl and HBr), aldehydes, alcohols, and other aliphatic and aromatic organic compounds. Little real information is available in the literature as to the additive, antagonistic, or synergistic effects of combustion products under conditions of reduced oxygen tension and increased temperature encountered during fire exposures.

Many of the newer synthetic materials, some widely used in industry and construction, as well as for waste containers, have vastly different flammability characteristics than the materials they replace (see Table 4.1). Better and more refined test protocols, such as the National Bureau of Standards test methods, are needed to more clearly define exactly what are the potential hazards of combustion products: these test results, converted into a total hazard index, will more clearly define how and where a material may be used and disposed of to achieve acceptable levels of fire safety.

In the article "Putting the Fire Toxicity Issues into Perspective," pages 16–18, Merrit M. Birky expresses concern over the highly emotional and polarizing battle which has emerged as fire deaths, and in particular, fire toxicity from inhalation of combustion products, becomes more widely recognized. By 1982, following several widely publicized fires where smoke inhalation was doubtlessly a factor, the subject of fire toxicity was being discussed openly, both in economic as well as in personal terms. Fire technology is a relatively new discipline in toxicology and in the fire service, and many questions will require years to answer. It is not clear that synthetics, yielding different combustion products than more common "natural" materials, such as cellulose, have been a major factor in recent fire-death history.

While research is underway at several locations on the real-world problem of smoke inhalation, such work requires time—months, years, even decades. Meanwhile, toxicity regulations and/or legislation to curb human fire losses are being considered by state and local jurisdictions. It is not clear that toxicity regulation will reduce life losses and injury. An overall fire hazard rating should incorporate toxicity as one parameter, but still consider the economic impact of the use of various alternative materials.

In another article from the magazine entitled "Toxicity Background Factors and the New York State Study," pages 19–21, Francis McGarry reviewed the 14 fire test methods considered by New York State, and noted that shortly after the Binghamton State Office Building fire (see Chap. 8) the New York State Legislature enacted the Chapter 552 of 1982, directing the Secretary of State to undertake a study of the toxicity of smoke and gases given off under various high-temperature and fire conditions. This study has now been completed, and recommends the University of Pittsburgh test method, and also that testing data filing should be instituted for materials used in the state.

In "Smoke Toxicity—A View from the Center for Fire Research," pages 22–23, Jack Snell notes that studies of the "Maryland Fire Fatality Study" between 1974 and 1977, involving detailed studies of 463 fire deaths, indicated that only 10% of these deaths were due to burns alone. CO and CO with HCN and/or heart disease were responsible for 64% of the fatalities. Since 1970, the Center for Fire Research of the National Bureau of Standards has developed test methods to detect or screen unusually hazardous materials before their widespread introduction and use. The relevance of these tests to the smoke-toxicity hazard in the "real world" remains to be established. Ongoing work is attempting to establish the relationship between smoke toxicity and the overall hazards of fire.

In "Control of Materials Based on Smoke Toxicity," pages 24–25, F. B. Clarke, III states that most of the literature on combustion product toxicology has appeared in the past 10 years. He feels that several unresolved problems remain:

1. Relating combustion conditions in a test to those in an actual fire.
2. Deciding how to measure toxicity (lethality vs. incapacitation).
3. Extrapolating animal data to humans.

No single test is likely to simulate more than a few of the wide variety of combustion conditions encountered in real life.

Since 90% of building fire deaths occur in residences, more attention should be given to residential housing, instead of to structures heavily impacted by building codes, such as commercial occupancies and places of public assembly. Building contents, especially furnishings, should have greater attention in this context.

Any approach should deal with burning behavior and then, if necessary, with toxicity. It is not realistic to attempt to make a big fire safe. Clarke feels that, in general, more positive impact would be achieved by controlling the rate of burning than by changing the composition of combustion products. Toxicity control should be approached with caution, and only attempted when there is a clear need for it.

In "Post Mortem-Protocol for Smoke Inhalation Fire Deaths," pages 26–29, Hugh Maguire reviews the Foundation for Fire Safety's plan for a more uniform information base, using standardized protocols for post-mortem studies. From the data thus obtained, many of the unresolved questions about fire toxicity and other effects of combustion may be resolved.

As noted previously, this series of magazine articles clearly shows the concern which is being expressed over the effects of hazardous and toxic chemicals, whether they are in the plant, the laboratory, the home, or in waste sites. The need for the protection of persons who are working with, in, or around waste chemicals, as well as in everyday contact with even known substances, is being recognized (see Chaps. 5 and 6).

Questions have been raised about the role of plastics in combustion toxicity. In the *Investor Quarterly 2,* 1983, pages 7 and 8, the B. F. Goodrich Company points out that the charge that PVC (polyvinylchloride) is a unique fire hazard has been made several times and each time has been refuted by clear-cut evidence to the contrary. Now the issue has been raised again, and over the past several months has received widespread news media attention. On one side are those, generally affiliated with manufacturers of competing materials, who contend that plastics, including PVC, give off deadly fumes when they burn and represent an unusual fire hazard.

Leading the attack is a large manufacturer of metal conduit for electrical wiring. This company, with assistance from the steel industry, is organizing a cleverly disguised smear campaign to arouse public alarm about PVC. Also involved is the

Foundation for Fire Safety, which purports to be an impartial scientific organization. However, as uncovered by an exposé in *Fortune* magazine earlier this year (1983), the foundation is actually funded by the metal-conduit manufacturer. Foundation officials frequently are interviewed by the media and participate at conferences and legislative hearings on fire safety. Plastics, they claim, are a major cause of fire deaths.

B. F. Goodrich has taken a leading role in countering these scare tactics. The company helped form the Vinyl Institute, a trade group spearheading a coordinated industry response to the attacks on PVC. Goodrich experts are making themselves available for news-media interviews and are working to educate legislative and regulatory authorities. The message is being heard. An article in *Fortune* on February 7, 1983, pages 68–74, entitled "The Dubious War on Plastic Pipe," exposed the effort to discredit PVC as a "well-documented example of unfair tactics in the marketplace" based on "an extraordinary outpouring of half-truths and misinformation." Also, through coordinated industry efforts, legislation that could have restricted the use of PVC in some states has been defeated or revised.

The key weapons in this counteroffensive are, for instance:

1. PVC helps keep fires from starting. Some recent well-publicized fires would not have started if PVC had been used in place of other materials.

2. *All* organic materials, including PVC, release toxic fumes when they burn. However, numerous tests show that the fumes released by PVC are not unusually harmful compared to other materials. PVC does emit hydrogen chloride when it burns; however, a Harvard University study of actual fires shows that a typical fire that includes PVC materials does not produce a harmful level of hydrogen chloride. The major threat is carbon monoxide, a combustion product of all organic material.

3. PVC is difficult to ignite. It must be heated to a very high temperature— over 700°F (371°C) before it will burn. It cannot be ignited by a match or a cigarette.

4. PVC cannot burn by itself. There must be an external flame present. Wood, on the other hand, ignites at less than 500°F (260°C) and burns independently.

5. When forced to burn, PVC burns slowly—90% slower than wood. This means PVC consumes oxygen, releases heat, and produces carbon monoxide more slowly than many other materials.

6. PVC has never been known to cause a fire death. Allegations that plastics killed persons in several well-publicized hotel fires are not supported by the fire and medical authorities who investigated these fires.

The main problem with ill-advised attempts to restrict the use of PVC and other plastics is that these efforts divert the public from attention to the *real* ways to reduce fire deaths—early detection and quick suppression of fire.

As a plastics-industry representative told a U.S. Senate Committee on July 19: "We know of no major United States fire disaster that has resulted in loss of life where modern building codes have been followed or where operating smoke detectors and sprinklers have been present.''

And, while the U.S. fire-death rate is still far too high, it is interesting to note that per capita deaths from fire decreased by 25% over the last decade, a period during which the use of plastics almost doubled.

(For more information on this issue, write to PVC Facts, Department 0122, WHB-5, B. F. Goodrich, 500 South Main Street, Akron, OH 44318.)

Open-pit burning was the commonly accepted disposal method for waste solvents and other chemicals until about 1960. The smoke and flame shown in Fig. 4.4 is typical of the pits which were used during that period, both as a disposal method of great utility for laboratory quantities of solvents, as well as for the training of personnel, including the plant fire brigade members, in the use of various extinguishing agents and techniques. Such open-pit burning has been outlawed by the Clean Air Act, as well as by the RCRA, but illustrates the concerns which we have expressed earlier in the chapter.

A very practical and workable alternative to open-pit burning is shown in Fig. 4.5. This device, a waste-solvent burner, was designed, fabricated, and operated using ''off-the-shelf'' oil burner pumps and piping, for disposal of laboratory liquids in 1961. It used kerosene or #2 fuel oil as the prime fuel, and LPG to start the

Figure 4.4. Open-pit burning, now illegal, was once used for disposal of waste solvents and for training of personnel.

Figure 4.5.   Control panel of burner for the safe disposal of laboratory quantities of waste solvents, assembled from "off-the-shelf" components. Safety can supplied kerosene for preheat after ignition of propane pilot. Waste solvents were fed from large tanks not seen on photo. Courtesy of General Electric Research Laboratory.

flame. Solvent was introduced from several tanks (not shown) at a rate of approximately 15 gal/hr. By making the proper selection of feed, even solvent mixtures containing chlorinated solvents could be burned without excessive air pollution. If more in-place devices of this type were available, the safe and effective disposal of solvents could be easily accomplished at a reasonable cost on-site, and would prevent the incident shown in Fig. 4.6.

Right to Know

The importance of knowing what is involved, from the viewpoint of fire, the environment, or control management as it attempts the risk/benefit analysis, has

Figure 4.6. Discharge of a flammable liquid into a storm sewer in Louisville, Kentucky. Object to right of underpass is automobile which was traveling on the street at the time of ignition.

been repeatedly debated in various legislative bodies as "right-to-know" laws were introduced. It is understandable that industry wishes to maintain proprietary information confidential, so it can be used to advantage in the marketplace. However, the recent action by the New Jersey state legislature is an interesting example of legislation that may well be considered by other states.

Senate No. 1670, Chapter 315, adopted by the legislature, was signed into law by the governor on August 29, 1983. This bill, known as the "Worker and Community Right to Know Act" has an interesting preamble:

> The legislature finds and declares that the proliferation of hazardous substances in the environment poses a growing threat to the public health, safety, and welfare; that the constantly increasing number and variety of hazardous substances, and the many routes of exposure to them, make it difficult and expensive to adequately monitor and detect any adverse health effects attributable thereto; that individuals themselves are often able to detect and thus minimize effects of exposure to hazardous substances if they are aware of the identity of the substances and the early symptoms of unsafe exposure; and that individuals have an inherent right to know the full range of the risks they face so that they can make reasoned decisions and take informed action concerning their employment and their living conditions.

> The legislature further declares that the extent of the toxic contamination of the air, water, and land in this State has caused a high degree of concern among its residents; and that much of this concern is needlessly aggravated by the unfamiliarity of these substances to residents.

The legislature therefore determines that it is in the public interest to establish a comprehensive program for the disclosure of information about hazardous substances in the workplace and the community, and to provide a procedure whereby residents of this State may gain access to this information.

The bill goes on to assign specific procedures for the workplace and environmental reporting of detailed information which would be useful.

One group which will doubtlessly benefit from this bill is the fire service. Having been involved in numerous chemical fires in New Jersey, such as the ones described previously, the emergency personnel of all services have a keen interest. One concession to industry in the bill is that only the five major ingredients of any chemical mixture need to be listed. Research and development laboratories will be covered in the bill as passed. The bill, under consideration in some form for almost 3 years, is effective 1 year from enactment (August 29, 1984) except that several departments charged with administration of the act are on immediate alert for implementation of the bill.

In another action, a federal appeals court has ruled that companies are now required to honor a union's request for health and safety data, as well as individual medical information, in all areas except "trade secrets." The U.S. Court of Appeals for the District of Columbia upheld the National Labor Relations Board orders against three companies in which local unions in 1977 had requested information about the industrial chemicals used at the work site and about the medical conditions of employees exposed to the substances.

Between the New Jersey law discussed above, and the court decision at the federal level on the medical data, the issue of "right to know" has come into the limelight.

BIBLIOGRAPHY

Center for Fire Research, "Further Development of a Test Method for the Assessment of the Acute Inhalation Toxicity of Combustion Products," NBSIR 82-2532, June 1982, National Bureau of Standards, Washington, D.C. (29 refs.).

"Dense Gas Dispersion," reprinted from *J. Hazard. Mater.* **6,** No. 1–2, 1–247, R. E. Britter and R. F. Griffiths (Eds.), Elsevier, Amsterdam, 1982.

Fawcett, H. H. and Wood, W. S. (Eds.), *Safety and Accident Prevention in Chemical Operations,* 2nd ed., Wiley–Interscience, New York, 1982. (See especially Chaps. 5–8, 17–19, 22, and 31.)

Federal Emergency Management Agency, National Emergency Training Center, Emmitsburg, MD 21727—residency programs in Incident Command, Advanced Incident Command, Fire/Arson Detection, Chemistry of Hazardous Materials, Hazardous Materials Tactical Considerations, Planning for a Hazardous Materials Incident, Emergency Medical Services, and other courses. (Contact the NETC, Emmitsburg, MD 21727 for schedule.)

Fire Protection Handbook, 15th Ed., National Fire Prevention Association, Batterymarch Square, Quincy, MA 02269, 1983.

Foster, H. D., *Disaster Planning: The Preservation of Life and Property,* Springer-Verlag, New York, 1982 (275 pp.).

Goodier, J. L., *Spill Prevention and Fail-Safe Engineering for Petroleum and Related Products,* Noyes Data Corp., Park Ridge, N.J. 1983.

Industrial Fire Hazards Handbook, 1st ed., 1979. (Available from the National Fire Protection Association, Batterymarch Square, Quincy, MA 02269.)

Landrock, A. H., *Handbook of Plastics Flammability and Combustion Toxicology: Principles, Materials, Testing, Safety and Smoke Inhalation Effects,* Noyes Data Corp., Park Ridge, N.J., 1983.

Melton, C. M. and Johnson, K., "Inspections, Drills Needed To Prevent and Deal with Fires," *Occup. Health Safety,* 38–43 (Jan. 1984) (4 refs.).

Norstrom, G. P., II, "Fire/Explosion Losses in the CPI," *Chem. Eng. Prog.,* 80–87 (August 1982); see also "An Insurer's Perspective of the Chemical Industry," in H. H. Fawcett and W. S. Wood (Eds.), *Safety and Accident Prevention in Chemical Operations,* 2nd ed., Wiley–Interscience, New York, 1982, Chap. 24, pp. 509–527.

"Prevention of Grain Elevator and Mill Explosions" (NMAB 367–2), National Accademy of Sciences, National Research Council, National Academy Press, Washington, D.C., 1982 (146 pp.).

"Report of the Committee on the Toxicity of the Products of Combustion to the Standards Council of the NFPA," *Fire Journal,* 21–27 (March 1983). (Report was prepared by Benjamin Clarke Associates for the National Fire Protection Association.)

Schumacher, M. M., *Landfill Methane Recovery,* Noyes Data Corp. Park Ridge, N.J. 1983.

5
Personal Protective Equipment

In addition to the respiratory protection/equipment, (see Chapter 6), several other types of personal protective equipment play an important role as "secondary" or back-up equipment where chemicals and other hazardous materials or substances are involved. It is generally agreed that proper engineering, adequate attention to safety and health practices, and education of both personnel and management so that most control is reduced to a standard operating procedure, carefully adhered to, cannot completely ensure employee or plant safety alone, since unusual incidents, spills, and failures, both human and mechanical, do in fact occur in the real world. Personal protective equipment is then called upon to provide that extra measure of operational safety and health.

The injurious effects of body, skin, respiratory, and eye contact with materials have been well documented elsewhere, but should be noted as ranging from acute irritation or injury, such as skin irritation, inflammation or lung involvement which might even result in pulmonary edema, or cardiac impairment, to more chronic injuries which may manifest themselves years later, including carcinogenic, tera-togenic, mutagenic, or reproduction-related effects. It is not clear even with the present knowledge whether one exposure to certain materials might induce highly undesirable effects years later.

Exposure or contacts may range from one-time splashes or inhalation, to con-tinuous wettings of liquids, or chronic inhalation of background "odors," as well as contact with solids. Clothing, as well as breathing protection, should be effective for more than a few minutes, to permit the wearer to retreat from the area, remove protective gear and clothing, and then thoroughly wash or flush with water and detergent. Under no circumstances should work clothing be worn off the site. Some provision must be made to ensure that a high level of personal hygiene, including, if possible, shower facilities and clothing change, is included. A portable facility, such as the shower and toilet facilities in a recreational vehicle or camper, is a possibility for on-site control of contamination, if operations are in an area remote from normal building facilities.

Protective clothing, including coveralls or protective suits, gloves, boots or rubber shoes, and face and head protection deserve critical attention. Unfortunately, no comprehensive performance criteria have been established or recognized for protective gear worn by persons in chemical-waste situations, and the present tables of protection provided by the makers and vendors of various equipment are less than adequate. Most are based on tests which may or may not apply to the exposure at hand. This lack of standards is especially disturbing, since, in most cases the waste site may be an unknown terrain. Until specifics are established, the unknown nature of the potential exposures must be recognized and respected (see Figs. 5.1–5.4).

A variety of materials from which protective clothing (suits and gloves) can be fabricated is available, from paper suits to plastic and other carefully designed suits and gloves, but each material fabrication and combination has limitations of use

Figure 5.1.   Air-supplied suit and hood of modern design, affording excellent overall protection if proper selection of plastic material is made. Courtesy of MSA.

Figure 5.2.   Light-weight inexpensive plastic/paper disposable clothing for temporary barrier, as used in a chemical carcinogen bioassay laboratory. Courtesy of H. H. Fawcett.

which must be considered. Normally, for routine occupational exposures, it would be recommended that any materials used in gloves, suits, or footwear to protect the wearer from contaminants should be tested for permeation before use. But when the nature of the contaminants and the degree or amount of exposures are unknown, the prudent wearer must rely on whatever test data are available for general classes of materials as analogous as possible to the match which is known. For example, if it is suspected that benzene may be encountered, good data on the permeability of benzene are available in the literature. For many other materials, and for combinations of chemicals which may be encountered in the field, corresponding data should be sought but often will be unavailable. In that case, the opinions of an industrial hygienist competent in this area should be sought and followed. The use of protective suits in sampling sewer systems is illustrated in Fig. 5.5 (see also Fig. 7.9).

Protective equipment, especially impervious clothing and respiratory protective devices, are not always given proper attention since, in addition to cost, they may be uncomfortable when worn for long periods of work. This discomfort will be especially severe when the devices and clothing are worn in hot or humid atmospheres. Various alternatives to minimize such discomfort are available, and may be considered, such as the use of air-supplied circulation to the suits to permit more nearly normal body heat and perspiration to be dissipated. In one model, a vortex air injector provides both internal air as well as cooling.

If an air supply is considered for breathing or cooling of an impervious suit, the

Figure 5.3.   Suit and hood affording more protection than clothing depicted in Fig. 5.2, including provision for self-contained breathing apparatus (SCBA) inside suit. Courtesy of MSA.

problems associated with air purity should be carefully analyzed. Diaphragm compressors or other devices which cannot generate carbon monoxide or excessive oil (from oil lubrication) should be noted. Cross connections to plant air supply (including the ever-present danger of some gas other than air being introduced into the system) must be considered. The quality of the air to the compressor or vortex must meet or exceed breathing air standards (see Chap. 6).

5.1.   HEAD PROTECTION

For chemical operations, especially in facilities with extensive overhead piping, tanks, columns, and related equipment which may occasionally leak or rupture,

Figure 5.4. Suit and hood similar to the ones in Fig. 5.3, with air-supply in addition to SCBA. Courtesy of MSA.

some form of head protection is clearly indicated. In addition, head protection may offer protection against impact, flying particles, and electric shock. If properly designed and engineered, head protection can protect the scalp, face, and neck as well as the head.

As noted by a recent study by the NIOSH, safety helmets are rigid headgear made of varying materials, such as polycarbonate plastic and even aluminum.[1-30] Two basic types are noted—full brimmed and brimless with peak. Both types are further divided into four classes:

Class A: Limited voltage resistance for general service.

Class B: High voltage resistance.

Class C: No electrical voltage protection (i.e., metallic).

Class D: Limited protection for fire-fighting use.

Figure 5.5. Clothing and breathing apparatus in use during underground systems sampling for dioxin at Love Canal. Note stand-by personnel with rope to assist if emergency develops (this is the "buddy system"). Courtesy of Geomet Technology, Inc.

Class A helmets are made without holes in the shell, except for those used for mounting suspensions or accessories, and have no metallic parts in contact with the head. They must pass voltage tests of 2200 V ac (rms) at 60 Hz for 1 min, with no more than 9-mA leakage. In addition, they should not burn at a rate greater than 3 in./min. After 24-hr immersion in water, the shell should absorb no more than 5% by weight. These helmets are designed to transmit a maximum average force of not more than 850 lb. Class A helmets do not weigh more than 15 oz, including suspension, but excluding winter liner and chin strap.

Class B helmets are designed specifically for use around electrical hazards, with the same impact resistance as Class A helmets. They pass voltage tests of 20,000 V ac (rms) at 60 Hz for 3 min, with no more than 9-mA leakage. In addition, no holes are allowed in the shell for any reason, and no metallic parts can be used in the helmet. The burning rate of the thinnest part is not greater than 3 in./min. After 24 hr in water, absorption of the shell will be no more than 0.5% by weight. The maximum weight of Class B helmets is 15.5 oz.[1,3a,b]

Class C metallic helmets do not afford the same degree of protection, but are preferred by many because of their lighter weight. They should not be used in the vicinity of electrical equipment, or near acids, alkali, and other substances that are corrosive to aluminum and alloys.

Brims are an optional feature of helmets. A brim completely around the helmet affords the most complete head, face, and back-of-neck protection. The brimless

Figure 5.6. Head protection should be comfortable, and compatible with other protection worn, as in this air-supplied respirator and "escape bottle." Courtesy of Scott.

with peak type of helmet may be used where a brim may be objectionable and the brimless may be equipped with lugs to support a face mask or face shield to provide face and eye protection.

The key to protection from impact is proper helmet suspension, since the suspension distributes and absorbs the impact force. Examples of suspensions include a compressible liner or a cradle formed by the crown straps and headband or a combination of the two. To function properly, the adjustment of the straps must keep the helmet a minimum distance of 1.25 in. above the head.

In order to keep the helmet from dislodging during normal use, various leather, fabric, and elastic chin straps are used. Helmet liners are optional equipment, often used in colder weather.

In recent times, the "bump" cap has been introduced to be worn by persons who are working in cramped quarters. In general, it is thin shelled and made of plastic. No specifications exist for these caps, and they are not a substitute for the helmets discussed above.

Another aspect of head protection which must be considered is the length of head hair in both males and females. This is especially critical where moving parts and machinery are involved. Hair nets or snoods are often specified, but worn only with reluctance by many workers. When combined with a cap, they have more appeal and hence increased utility. In areas where sparks or hot metals are encountered, such as in welding or cutting operations, the nets or caps should be flame-resistant and have a visor sufficiently long to afford protection from the exposures coming from above on an angle.

A wide variety of plastics, fiberglass, and other materials is available for helmet construction, and many are available in colors. The use of colors often contribute to immediate identification of various crafts, such as millwrights, electricians, pipefitters, and also is a type of security clearance to the worker, especially in a large facility. The wearing of the employee's name on the hat is also a useful incentive for each person to keep his or her hat in a proper condition of cleanliness. Earmuffs or ear protectors are often integrated into hats, and, when properly adjusted, are sufficiently comfortable for long periods of wearing.

It is important that the headgear be compatible with other personal protective equipment (see Fig. 5.6).

5.2. HEARING PROTECTION

The question of how much exposure to certain levels of noise of certain frequencies causes hearing loss is not completely resolved. However, we live in a world which seems to become noisier, as heavy traffic, subways, radios, televisions, stereos, public address systems, outboard motors, power lawn mowers, jet engines, rocket engines, helicopters, and supersonic booms compete for our attention. Within the chemical industry, it is possible to note agitators, grinders, mixers, pulverizers, compressors, fans, blowers, and other equipment which are noisier than the generally accepted standard of 90 dBA. At which point for certain periods of time the noise becomes a problem can only be determined by measurements, analyzing both the frequency and the length of exposure to the noise. Although industrial hygiene engineers usually have the qualifications and equipment for making surveys, frequently an acoustical engineer is required to make the final determinations and recommendations to the plant physician or occupational health nurse. The medical department may elect to institute a regular program of audiometric testing, during which the employee is tested at regular intervals, such as yearly or semiannually, for his or her response to various frequency tones which correspond to normal speech. The records of such a test then constitute an important part of the employee's overall medical records, and can be referred to in the future to determine if a tendency toward hearing loss is occurring, and whether it is from occupational or nonoccupational causes. Since disco music is often in the range of 100–120 dBA, and exposure may be for significant times after work, it may be difficult to pinpoint the exact cause of hearing impairment.

Once a program is agreed upon, the medical department may recommend the

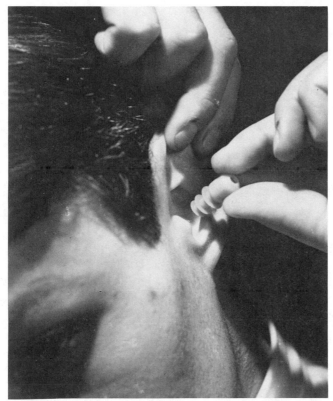

Figure 5.7. Ear plugs, when required, should be carefully fitted by medical personnel to ensure proper placement in the ear.

proper fitting of protective devices, such as earplugs, earmuffs, helmets, or combinations of these, so the individual is matched for the specific job requirements. It is vital that the individual understand the importance of this program to him or her as an individual, and that he or she actually cooperate in the fullest. Earplugs, which are the simplest and least expensive control, have only limited ability to attenuate sound and must be carefully fitted, as illustrated in Fig. 5.7; for the higher-sound levels, earplugs must be supplemented by muffs, or helmets with built-in muffs (see Fig. 5.8). To facilitate voice communications, built-in earphones and attached microphones are available to permit the use of telephones or two-way radio equipment.

5.3. SAFETY FOOTWEAR

Safety shoes and boots have become widely accepted in the chemical industry, since the intrinsic value of the steel safety cap, which weighs little more than an ounce, has been demonstrated many times. In addition, safety shoes and boots, in

Figure 5.8.   Ear muffs can be adapted to earphones for radio communications. Courtesy of Norton.

styles both for women and men, are attractive, and represent an excellent value. Many companies encourage the use of safety shoes by giving partial or total credit for shoe purchase, and the shoe vendors often have mobile shoe vans or trucks which make on-the-job fitting most convenient.

According to the NIOSH, protective footwear may be classified into five categories:

1. Safety-toe shoes (the most commonly accepted).
2. Conductive shoes (to leak off static electricity).
3. Foundry or molders' shoes.
4. Explosive-operations shoes (nonsparking).
5. Electrical hazard shoes.

Although not a class as such, acid shoes, that is, safety shoes or boots made of materials such as neoprene which resist attack by acids, alkali, and other corrosives, are important to acid-handling operations, including tank car loading and discharging.

OSHA requires the use of safety-toe shoes for persons who work with heavy material, since the shoes afford protection against impact and rolling objects, and against the hazard of inadvertently striking sharp sheet metal. Safety-toe shoes have been divided into three groups, namely 75, 50, and 30, which represent the minimum requirements for both compression and impact. The usual material of construction for the safety toe is iron. For additional protection against extremely heavy impacts, metatarsal (or over-feet) guards may be used in conjunction with safety-toe shoes. These guards are made of heavy gauge metal, flanged, and corrugated to protect the feet to the ankles.

In locations where there is a potential for a fire or explosion, such as the liquefaction of hydrogen or methane, or in the handling of highly volatile substances such as ether, conductive shoes should be worn. These shoes are engineered to dissipate static charges, reducing the possibility of static spark, when used in connection with a conductive flooring, and the shoes and floors are properly maintained to ensure real conductance. There has been much controversy as to whether or not a static discharge from a human would institute or cause ignition; the evidence is sufficiently strong that no prudent person can argue against the use of every precaution.[5] Periodic tests should be performed, using proper equipment, to ensure that the maximum allowable resistance of 450,000 Ω is not exceeded.

Foundry shoes protect against molten metal splashes, and are made so they may be removed easily and rapidly if a spill occurs. The tops of the shoes should be closed by the pants leg, spats, or leggings to preclude splashes entering from the top.

In hazardous locations, such as those areas where the floors are not conductive and grounded, and in which highly flammable and volatile gases and liquids are handled, such as gasoline tanks, nonsparking shoes should be worn. These shoes do not have conductive soles, and have nonferrous eyelets and nails. The metal toe boxes are coated with nonferrous metal, such as zinc or aluminum.

Electrical hazards, such as those presented by the potential for contact with an electrical current running from the point of contact to ground, may be minimized by the wearing of electrical shoes. These shoes contain no metal, except for the toe box, which is insulated from the remainder of the shoe. Dampness or significant wear will definitely decrease the protection provided by these shoes.

5.4. HAND PROTECTION (GLOVES)

Although gloves were discussed briefly in a previous section of this chapter, the extensive use of various types of gloves, for protection against a variety of insults including chemical, should be mentioned. In addition to toxics, gloves are available for protection against abrasions, cuts, heat, and cold.

Two basic methods are used in manufacturing liquid-proof gloves—the latex-dipped process and the cement-dipped process. In the former, particles of rubber, suspended in water, are coagulated onto a glove form. Since numerous tiny air pockets may develop that have resulted from the incomplete contact of rubber particles with the form, the latex-dipped gloves have a higher penetration or permeation rate than do the cement-dipped gloves. In the cement-dipped process, the rubber is dissolved in an organic solvent before it is deposited on the form. This permits a more complete joining of rubber particles resulting in fewer air pockets and greater resistance to permeation.[6]

NIOSH has noted other factors that affect the rates of permeation, namely, glove thickness and solvent concentration. (To this should be added the type of solvent, since some materials may be highly resistant to a solvent, while others are not.) Permeation rates are indirectly proportional to glove thickness and directly proportional to solvent concentration.[6] Naturally, time of contact and area involved should also be considered.

As noted in the section on protective clothing, the first consideration in selection of gloves is to ascertain, insofar as possible, the exact nature of the substances to be encountered, and to note any possible reactions which might occur between the solvents and the glove materials. Chemical resistance charts are made available to prospective buyers by the glove manufacturers. These charts should be examined critically, and the test data supporting the classification should be requested from

Figure 5.9.   Long-sleeve gloves provide protection for hazardous exposures, if selected from the proper material and maintained properly. In this illustration, impervious long sleeves on shirt would be desirable. Courtesy of Norton.

the laboratory which developed the chart. (As noted previously, no generally agreed test method exists for gloves or other protective clothing.) Since most gloves are supplied in several different thicknesses, the choice must be made as to most complete protection versus flexibility. In general, the thickest glove practical should be specified, if it is not too stiff and will not introduce an additional hazard into the work. The length of the glove is important, since it should be sufficiently long to protect the forearms as well as hands. If liquids are handled, long gloves permit the wearer to turn down the top of the gloves to form cuffs, thereby preventing the liquid from passage down the arms when the hands are raised[6] (see Fig. 5.9).

5.5. SAFETY BELTS, LANYARDS, LIFELINES, AND SAFETY NETS

Where work is to be performed at significant heights, safety belts and lifelines should be used. During normal operations, such as hoisting or lowering the wearer, or providing him or her with steady support while at work, the belt or lifeline may be considered part of the routine equipment and procedure for the job. (This includes the lowering of personnel into utility holes or tanks where no ladder is available or cannot be used for some reason.) Under these normal operations, comparatively mild stresses are applied to the belt, and are usually less than the total static weight of the wearer. In emergency use, however, such as preventing the wearer from falling further than the slack in the line, the belt or line is subjected to impact loading, which can amount to many times the weight of the user. The amount of impact force developed in stopping a fall depends on three factors:

1. The weight of the user.
2. The distance of the fall.
3. The speed of deceleration or stopping.

To limit the distance of the fall, the wearer should never tie off below waist level. Some shock absorber or decelerating device (such as a spring or elastic rope) should be incorporated into the system to bring the fall to a gradual stop. The impact load on the equipment and the wearer will thus be reduced.

Four classes of safety belts have been recognized:

Class I—Body belts for limited movement and positioning to restrict the worker to a safe area in order to prevent a fall.

Class II—Chest harnesses used where freedom of movement is paramount, and only limited fall is possible. These are not recommended for vertical free-fall hazard situations.

Class III—Body harnesses used when the worker must move about at dangerous heights. In the event of a fall, the harness distributes the impact force over a wide body area, reducing the injury potential.

Class IV—Suspension belts used when working from a fixed surface. The user is completely supported by the suspension belt. These are used in shipboard and bridge painting and maintenance, stack maintenance, and tree trimming.

Of the four classes, the body harnesses are preferred for many operations, since they more completely absorb impact and will keep the wearer upright during a fall. When choosing belts and harnesses for protection and selecting personnel for the job, all aspects of the hazard should be recognized—the possible distance to fall, the limitations against restricted mobility, and the health and stability (both physically and emotionally) of the wearer.

To judge the safety of a belt, the NIOSH recommends several criteria: The belt should be of sufficient strength to withstand the maximum possible free fall of the wearer. The lanyard of the safety belt should be a minimum of 0.5-in. nylon (or equivalent) with a maximum length that limits a fall to 6 ft or less. The rope should have a nominal breaking strength of 5400 lb. Also, the belt should be equipped with some form of shock absorber to limit the impact loading. The stopping distance of the belt should be such that it will prevent the user from striking some dangerous obstruction before the fall is arrested. There must also be a sufficient safety margin on all of the above criteria to cover all unknowns, including the weight and physical condition of the wearer, the distance of the fall, the distance to dangerous obstructions, variations in the strength or elasticity of materials, and deterioration of the materials due to wear or other causes.[1,2]

Lifelines, which attach the user of a safety belt to an anchorage, such as below surface or window washing and similar maintenance, should be secured above the point of operation. This positioning will limit the fall of the wearer, thus reducing the impact loading. All lifelines should be capable of supporting a minimum deadweight of 5400 lb. The most important criteria for lifelines are their shock-loading strength and energy-absorption ability. Shock absorbers should be used to assist the lifeline in these critical areas. Lifelines should be secured so the wearer is subjected to as little shock as possible. This will limit the free fall.

Numerous materials are available for lifelines including nylon, dacron, and manila. Each has specific characteristics, and lifelines should be selected based on the specific situation in which they will be used.

Safety nets represent another approach to minimization of fall injuries. Safety nets should be used wherever the use of safety belts and lifelines is impractical or infeasible where the protection from falls is important.

Safety nets are usually designed for specific purposes and are therefore custom-made from various natural and synthetic materials. These materials are used in the form of rope or webbing and in numerous combinations of rope or webbing in various mesh sizes. Each net and section of net should have boarder ropes of the same material, but of larger diameter. The Associated General Contractors and the U.S. Corps of Engineers have established minimum sizes for safety net materials, as well as specific performance standards for nets of various sizes and dimensions.[7]

Where the use of safety nets is considered, extra care should be exercised to arrange the nets so sufficient clearance exists to prevent the nets from contact with surfaces or structures below or to each side when the anticipated impact load is applied. When using safety nets near electrical power lines, this factor becomes especially critical.[7]

When more than one net is employed and are joined to form a larger net, they should be laced or otherwise secured so they perform properly. For all nets, perimeter suspension systems should be designed and installed in such a manner that the suspension points are either level or slope toward the building so a rebounding load will be directed into a protected area. Perimeter nets should always follow the work upward as it progresses, and should never be more than 25 ft below the working level, and if possible, closer.

In order to ensure that nets perform as anticipated, daily inspections are essential. Inspections should be made prior to and after installation, after any alterations, and after impact loading occurs. Recommended test procedures are available.[8] All defects should be repaired properly and promptly.

5.6. SUMMARY

Selection, regular use, care, cleaning, and maintenance of personal protective equipment must be based on complete understanding of the hazards to which the worker is exposed. For example, a worker exposed to electrical hazards as well as foot and head injuries should be equipped with a Class B, electrical-resistant safety helmet, and with special electrical hazard shoes that have no soles of flexible metal. In addition, the worker must be adequately and periodically instructed in proper use of the equipment, the importance of the equipment to his or her health and safety on the job, and the proper methods of adequate maintenance. He or she must understand that a protective device for one job is not always adequate for another, and hence he or she should not loan equipment to another worker. A paint-spray respirator, for example, is not effective against chemical acid gases. Gloves which may be adequate protection for paint dipping may not be proper protection for wearing around a trichloroethylene degreaser (see Chap. 6).

No protective equipment is without some disadvantage, but should be as comfortable as possible, and properly fitted and maintained to encourage use. The equipment should not increase or introduce additional hazards to the job. Visibility, weight, and appearance are often factors in the "real-world" use. A full-brimmed hard hat used in close quarters where mobility is limited, for example, may obstruct the wearer's vision, and make him or her prone to other hazards. We have deliberately omitted eye protection from this chapter in spite of its great importance. (See J. Nichols, "Eye Safety in Chemical Operations," in H. H. Fawcett and W. S. Wood (Eds.), *Safety and Accident Prevention in Chemical Operations,* 2nd ed., Wiley–Interscience, New York, 1982, Chap. 26, pp. 573–581.)

Unless the protective devices are accepted for their intrinsic value, *and actually used,* a false sense of security, which may result in serious problems, will result from an inadequate program.

REFERENCES

1. J. Chapman and G. W. Pearson, *Safety Information Profile, Personal Protective Equipment Usage in Industry,* National Institute for Occupational Safety and Health, Morgantown, W. Va., 1980.

2. *Accident Prevention Manual for Industrial Operations,* 8th ed., National Safety Council, Chicago, Ill., 1977.

3a. "Safety Requirements for Industrial Head Protection," ANSI Z89.1-1969, American National Standards Institute, New York, 1969.

  b. W. I. Cook and D. W. Groce, "Report on Tests of Class B Industrial Helmets," National Institute for Occupational Safety and Health, Cincinnati, Ohio, 1975.

4. "Safety Requirements for Industrial Protective Helmets for Electrical Workers, Class B," ANSI Z89.2-1971, American National Standards Institute, New York, 1971.

5. R. W. Johnson, "Ignition of Flammable Vapors by Human Electrostatic Discharges," Alleghany Ballistics Laboratory, Hercules, Inc., Cumberland, Md., 1980.

6. W. H. Figard, "Intensifying the Efforts of Proper Glove Selection," *Occup. Health Safety* **49,** No. 7, 30, 42, 43 (1980).

7. Construction Hazards Committee, "Construction Management Bulletin CM-4.0, Safety Nets," American Insurance Association, New York, 1972.

8. "Safety Nets, Data Sheet 608," National Safety Council, Chicago, Ill., 1967.

BIBLIOGRAPHY

Ahmed, I., "Protective Clothing and Skin Contact—A Bibliography" (first draft 6-23-80), National Safety Council, Chicago, Ill., 1980.

Anon., New Data Announced on Safety Glove Permeation," *Update,* Premier Issue, 1–8 (May 1982). (Available from Lab Safety Supply, P.O. Box 1368, Janesville, WI 53547.)

Blackman, W. C., Jr., "Enforcement and Safety Procedures for Evaluation of Hazardous Waste Disposal Sites," *Management of Uncontrolled Hazardous Waste Sites,* U.S. EPA National Conference, October 15–17, 1980, Washington, D.C., pp. 91–106.

Coletta, G. C., "Development of Performance Criteria for Protective Clothing Use Against Carcinogenic Liquids," NIOSH Technical Report, DHHS (NIOSH), October 1978.

Coplan, M. J., and Lopatin, G., "Preparation of Activated Carbon-Filled Microporous Hollow Multifilament: A Summary Report," AD-A067663, 1979. N.T.I.S.

"Dressing Right for Safety," *Natl. Saf. News,* 104–107 (March 1980).

Federick, E. B., and Henry, M. C., "A Study of Seam Leakage in Coated Fabrics: Summary Report," AD 708-874, August 27, 1979, National Technical Information Center, Springfield, Virginia.

Fourt, L., "Heat and Moisture Transfer through Fabrics: Biophysics of Clothing," AD 684-949, November 1959.

Hart, John A. H., "Cellular Resin Foams Resistant to the Passage of Noxious Chemicals in Liquid and Vapor Forms," Canadian Patent No. 878560, May 4, 1979.

Houston, Davis E., "The State of the Art—Safety Wearing Apparel," *Natl. Saf. News,* 115–120 (March 1978).

"If They Have to Wear It, Tell Them to Wear It Right," *Natl. Saf. News,* 136–142 (March 1970).

Johnson, K. E., and Lowish, M. D., "Protection Should Fit Worker, Job," *Occup. Health Safety* **52,** No. 8, 38–43 (August 1983).

Lynch, P., "Matching Protective Clothing to Job Hazards," *Occup. Health Safety,* 30–34 (January 1982).

Michael, P. L., and Bienvenue, G. R., "Hearing Protector Performance—An Update," *Am. Ind. Hyg. Assoc. J.* **41,** No. 8, 542–546 (August 1980).

Mihal, C. P., Jr., "Effect of Heat Stress on Physiological Factors for Industrial Workers Performing Routine Work and Wearing Impermeable Vapor-Barrier Clothing," *Am. Ind. Hyg. Assoc. J.* **42,** No. 2, 97–103 (February 1981).

Morrow, R. W., and Hamilton, J. H., "MOCA Permeation of Protective Clothing," Oak Ridge Y-12 Plant, Department of Energy, Y-DK-109, November 7, 1979.

Moshkovich, L. G., and Aleksandrova, T. M., "New Semiwool Fabric with Acid-Resistant Finishing for Protective Clothing." *Tekst. Promst. (Moscow)* **78,** No. 2, 52–53 (1979).

Mychko, A. A., Efremov, V. A., and Andkhin, V. V., "Methods for Determining the Acid Permeability of Special Materials," *Iv. Vyssh. Uchen, Zaved. Tekhnol. Legk. Promsti* **78,** No. 21 (6), 8–11 (1980).

Oliver, T. N., "How Protective Clothing Industry Is Coping." *Natl. Saf. News,* 105–110 (July 1974).

"Programming Personal Protection—Everybody Needs Protection," *Natl. Saf. News,* 104 106 (November 1975).

Royster, L. H., "An Evaluation of the Effectiveness of Two Different Insert Types of Ear Protection in Preventing TTS in an Industrial Environment," *Am. Ind. Hyg. Assoc. J.* **41,** No. 3, 161–169 (1980).

Sansone, E. B., and Tewari, Y. B., "The Permeability of Protective Clothing Materials to Benzene Vapor," *Am. Ind. Hyg. Assoc. J.* **41,** No. 3, 170–174 (1980).

Sansone, E. B., and Tewari, Y. B., "Differences in the Extent of Solvent Penetration Through Natural and Nitrile Gloves from Various Manufacturers," *Am. Ind. Hyg. Assoc. J.* **41,** No. 7, 527 (July 1980).

Schnabel, G. A., Christofano, E. E., and Harrington, W. H., "Safety and Industrial Hygiene During Investigations of Uncontrolled Waste Disposal Sites," *Management of Uncontrolled Hazardous Waste Sites,* U.S. EPA National Conference, October 15–17, 1980, Washington, D.C., pp. 107–110.

"Standards Are in Store for Protective Clothing," *Chem. Eng.* **63,** 76–79 (April 21, 1980).

Tabor, M., "Women in Coal Mines: PPE, Where Are You?" *Occup. Health Safety,* 22–26, (January 1982) (A Safety and Health Personal Protective Equipment Checklist for Women is on p. 25.)

Vo-Dihn, T., and Gammage, R. B., "The Lightpipe Luminoscope for Monitoring Occupational Skin Contamination," *Am. Ind. Hyg. Assoc. J.* **42,** No. 2, 112–120 (February 1981).

Votta, F., Jr., "Flow of Heat and Vapor through Composite Perm-Selective Membranes," AD 671–681, January 1968, National Technical Information Center, Springfield, Virginia.

Weeks, R. W., Jr., and McLeod, M. J., "Permeation of Protective Garment Material by Liquid Benzene," LA-8164-MS, Los Alamos Scientific Laboratory, Los Alamos, N.M., December 1979.

Williams, J. R., "Permeation of Glove Materials by Physiologically Harmful Chemicals," *Am. Ind. Hyg. Assoc. J.* **40,** No. 10, 877–882 (1979).

Williams, John R., "Chemical Permeation of Protective Clothing," *Am. Ind. Hyg. Assoc. J.* **41,** No. 12, 884–887 (December 1980).

Head Protection

"Controversy Swirls Around Head Protection," *Occup. Hazards* **18,** 22–26, 163 (January 1981).

Head Protection, ANSI Z89.1-1981, American National Standards Institute, New York, 1981.

Noise and Hearing Protection

Burns, W., and Robinson, D. W., "Hearing and Noise in Industry," Department of Health and Social Security, Her Majesty's Stationery Office, London, 1970.

"Compendium of Materials for Noise Control," Contract No. HSM 99-72-99, DHEW Publication No. (NIOSH) 75-165, June 1975.

Gasaway, D. C., "Safety in Numbers Misleading in Assessing Auditory Risk Criteria," *Occup. Health Safety,* 58–66 (January 1984) (8 refs.).

Heffler, A. J., "Audiometry Does Not Equal Hearing Conservation," *Occup. Health Safety,* 38–40, 42, 45 (March 1982).

Hynes, G., *Occup. Health Safety,* 38–40, 42, 45 (March 1982).

Hynes, G., "To Buy or Not to Buy: The Dosimeter Vs. Sound Level Meter Question," "To Buy or Not to Buy: The Dosimeter Vs. Sound Level Meter Question," *Occup. Health Safety,* 18–25, 57–60 (March 1982).

"Industrial Noise Manual," American Industrial Hygiene Association, Akron, Ohio, 1979.

Lund, A. O., "Noise Control Enclosures for Industrial Equipment," *Am. Ind. Hyg. Assoc. J.* **40**, No. 11, 961–969 (1979).

Michael, P. L., "Industrial Noise and Conservation of Hearing," in G. D. Clayton and F. E. Clayton (Eds.) *Patty's Industrial Hygiene and Toxicology,* Vol. 1, 3rd rev. ed., Wiley–Interscience, New York, 1978.

Moselhi, M., "A Six-Year Follow Up Study for Evaluation of the 85 dBA Safe Criterion for Noise Exposure," *Am. Ind. Hyg. Assoc. J.* **40**, No. 5, 424–426 (1979).

"Noise Control: A Guide for Workers and Employers, Occupational Safety and Health Administration," U.S. Department of Labor, Washington, D.C., 1980.

Reischl, U., "Fire Fighter Noise Exposure," *Am. Ind. Hyg. Assoc. J.* **40**, No. 6, 482–489 (1979).

Royster, L. H., and Thomas, W. G., "Age Effect Hearing Levels for a White Nonindustrial Noise Exposed Population (NINEP) and Their Use in Evaluating Industrial Hearing Conservation Programs," *Am. Ind. Hyg. Assoc. J.* **40**, No. 6, 504–511 (1979).

"Safety Guidelines and OSHA Summaries, Noise and Vibration Analysis, Noise and Vibration Reduction, Hearing Protection," in *Best's Safety Directory, Industrial Safety, Hygiene, Security,* Vol. 1, A. M. Best Co., Oldwick, N.J., 1979, Chap. 5, pp. 401–459.

Sataloff, J., "Noise Regulation: Resounding Need for Joint Cooperation," *Occup. Health Safety,* 10–11 (January 1982).

Sataloff, J., and Sataloff, R. T., "Ear Protectors Vs. Intense Impact Noise," *Occup. Health Safety,* **52**, No. 8, 25–28 (August 1983).

Schmidek, M. E., "Survey of Hearing Conservation Programs in Industry," DHEW Publication No. (NIOSH) 75-178, June 1975.

Thunder, T. D., and Lankford, J. E., "Relative Ear Protector Performance in High Vs. Low Sound Levels," *Am. Ind. Hyg. Assoc. J.* **40**, No. 12, 1023–1029 (1979).

"Were Noise Controls 'Technologically Feasible' "? *Occup. Hazards,* 37–38 (January 1981). (See also p. 160 of same issue.)

Clothing

Ahmed, I., "Criteria Used by Companies in Choice of Chemical Protective Clothing (Results of a Survey)," National Safety Council, Chicago, Ill., 1980.

Arons, G. N., and MacNair, R. N., "Laminated, Highly Absorbent Activated Carbon Fabric," U.S. Patent No. 46733, June 7, 1979.

"ASTM Standard Test Method for Resistance of Protective Clothing Materials to Permeation by Hazardous Liquid Chemicals, Draft of Test Method," ASTM D WXYZ-80, American Society for Testing and Materials, Philadelphia, Pennsylvania, 1980.

Berger, M. R., "Safety Clothing—A Matter of Personal Protection," *Natl. Saf. News,* 63–67 (September 1976).

Bienvenue, G. R., and Michael, P. L., "Permanent Effects of Noise Exposure on Results of a Battery of Hearing Tests," *Am. Ind. Hyg. Assoc. J.* **41**, No. 8, 535–541 (August 1980).

Bosserman, M. W., "How to Test Chemical-Resistance of Protective Clothing," *Natl. Saf. News,* 51–53 (September 1979).

"Test Method for Resistance of Protective Clothing Materials to Permeation by Hazardous Liquid Chemicals (Part 46, 1982)," ASTMF 739-81, American Society for Testing and Materials, Philadelphia, Pennsylvania.

6

Respiratory Protective Equipment

Humans, plants, and animals are addicted to "air," a unique mixture of gases that is the least appreciated but most essential and valuable basic resource of the planet earth. Air has been recognized as elemental by the Ancient Greeks, who held air, along with water, earth, and fire, as basic to life. Without air, life is measured in minutes. As the term is commonly used, air is approximately 78% by volume nitrogen, 21% by volume oxygen, and less than 1% by volume of other gases including helium, argon, neon, krypton, xenon, and usually small amounts of carbon monoxide, carbon dioxide, sulfur dioxide, oxides of nitrogen, ozone, and moisture as well as particulates (aerosols).

We note these ingredients which normally comprise air to stress that even air fully suitable for human consumption usually contains varying concentrations of many substances in addition to the essential life-sustaining oxygen.

In the real world, air is constantly being modified or changed by numerous agents, processes, and activities. Included in these are natural forces such as precipitation which contributes moisture; the winds which are carriers of dusts, pollen, seeds, spores, salt, and other particulates; changes or stagnation in atmospheric pressures which contribute to air movement or lack of it (stagnation or lapse conditions); lightning which contributes ozone and oxides of nitrogen; volcanic activity (such as the 1980 eruption of Mt. St. Helens in Washington State) which contributes sulfur dioxide and ash; decay of radioactive nuclides of long half-life which contribute radon gas and other "daughter" products; fire (from both natural and person-related causes) which contributes carbon dioxide, carbon monoxide, and particulates in smoke; biological processes, such as fermentation and decay, which contribute carbon dioxide, and occasionally, carbon monoxide, hydrogen sulfide, and methane by action of bacteria, enzymes, and other agents; and the sun which contributes ultraviolet, visible, and infrared radiation. Solar radiation is an important ingredient of this complex mixture by providing light and heat, the energy for plant photosynthesis (important in the oxygen–carbon dioxide balance), as well as catalyzing the action of oxidizers, such as ozone and oxides of nitrogen, on hydrocarbon gases and vapors with particulates to synthesize "smog" (smoke and organic matter). Gases dissolved in the water of natural springs and spas evolve carbon dioxide (carbonated water) and hydrogen sulfide (mislabeled "mineral waters or radium waters" in some locations).

As mentioned above, another gas in air which is receiving considerable attention is radon, an inert radioactive gas, formed from naturally occurring uranium and radium. It can enter buildings through stone and concrete, as well as through the water supply. Radon is suspect as a causative agent in lung and stomach cancer, which have been studied in connection with uranium mining. Some homes, especially "winterized" homes, have significant concentrations of radon.

Radon can enter a water supply from rocks and soil in contact with well water or reservoir water. Well water is frequently higher in radon compared to public water. Levels of 5000 to 10,000 pCi/liter of radon in water can be cause for concern.

A commercial radon-water-measurement service is available at the University of Maine, calibrated by exposure of detectors to known concentrations of radon in water.[1a,b]

Another unusual entry to the list of components in air is bromine. A repeatable springtime burst occurs in solid (aerosol) and gas-phase bromine in arctic air (troposphere). At Point Barrow, Alaska, the aerosol fraction of tropospheric samples averaged about 5 ppt. During the springtime, particulate bromine levels exceeded 82 ppt. Gas-phase bromine averaged 7 ppt, over the spring bloom, and total bromine in arctic air jumped to about 133 ppt. More investigations are underway.[1d]

Temperature extremes, in which air is too cold or too hot to breathe without tempering, may also contribute to respiratory stress. Table 6.1 shows the composition of unpolluted air.[1,1a–e]

To this limited inventory of natural sources which contribute to atmospheric contamination must be added products from the many activities of humans. For our purposes, it may suffice to state that most humans can tolerate a significant deviation from "pure breathing air," regardless of the source of the contamination, since most humans have amazing powers and the ability to adjust or to tolerate "insults." It is only when the summation of the various pollutants in the breathing zone reaches an *intolerable* level that human life is threatened by respiratory hazards, whether of artificial or natural origin.[4] Serious effects may be observed in seconds from a few breaths of some gases, such as arsine or phosgene, or observed decades after chronic exposures to an aerosol containing bis-dichlorodiethyl ether or to asbestos.[5]

Aerosols, consisting of solids or liquids suspended in air, are characterized in Table 6.1. It is recognized that the human system in generally good health has remarkable tolerance or ability to cope with assaults from inhalation insults, but this tolerance has an upper limit beyond which injury will eventually result. Hence the term "threshold" is applied to limits above which airborne contaminants may be expected to produce injury or ill effects of some type. However, this threshold is not a go no-go concept. For a small but very real percentage of persons who constitute a high-risk population, including the very young, the elderly, and persons with severe respiratory diseases, such as bronchial asthma, chronic bronchitis,

TABLE 6.1. The Gaseous Composition of Unpolluted Air (Dry Basis)

	Percent by Volume	Parts per Million by Volume
Nitrogen	78.09	780,900
Oxygen	20.94	209,400
Water	—	—
Argon	0.93	93,001
Carbon dioxide	0.03	325
Neon	Trace	18
Helium	Trace	5.2
Methane	Trace	1.0—1.2
Krypton	Trace	1.0
Nitrous oxide	Trace	0.5
Hydrogen	Trace	0.5
Xenon	Trace	0.08
Organic vapors	Trace	~0.02

Source. A. C. Stern et al. *Fundamentals of Air Pollution,* Academic Press, New York, 1973, p. 21.

pulmonary emphysema, or pneumonia, exposures which may be permissible or acceptable to most healthy persons may have serious effects.

Respiratory hazards have been studied over a century, especially in Germany and England, but even today much is not completely known.[6] We are limited in our understanding about the synergistic or combined effects of air contaminants. We do know that several substances have enhanced toxicity when inhaled with a "carrier." However, studies of "real-world" air, including the concentrations of the various gases, vapors, and particulates encountered in daily exposures, especially when aggravated by smoking, alcohol, and physiologically active aerosol drugs, are seldom directly applicable to an individual because our data base is usually too incomplete.[7-9]

Due to the importance of aerosols to respiratory studies, working definitions of certain terms may be useful:

PARTICLES: Any minute piece of solid or liquid. Many particles that are important in studies of air pollution are unstable—they can change or even disappear on contact with a surface. Examples are a raindrop striking a surface and coalescing, a loose aggregate of carbon black disintegrating on contact with a surface, and an ion losing its charge after contact with a surface or an oppositely charged particle.

AEROSOLS: A mixture of gases and particles that exhibit some stability in a gravitational field. In atmospheric aerosols, this gravitational stability excludes particles[10,11] with a diameter greater than several hundred micrometers. See Table 6.2 for names and characteristics of aerosol particles.

Under the Federal Clean Air Act, standards have been set for six air pollutants: carbon monoxide, sulfur dioxide, nitrogen dioxide, particulate matter, gaseous

TABLE 6.2.   Names and Characteristics of Aerosol Particles

Name	Unique Physical Characteristics	Effects	Origin	Predominant Size Range (μm)
Coarse particle			Mechanical process	>2
Fine particle			Condensation	<2
Dust	Solid	Nuisance and ice nuclei	Mechanical dispersion	>1
Smoke	Solid or liquid	Health and visibility	Condensation	<1
Fume	Solid	Health and visibility	Condensation	<1
Fog	Water droplets	Visibility reduction	Condensation	2–30
Mist	Water droplets	Visibility reduction and cleansing air	Condensation or atomization	5–1000
Haze	Exists at lower RH than fog—hygroscopic	Visibility reduction		<1
Aitken or condensation nuclei (CN)		Nuclei for condensation at supersaturation <300%	Combustion and atmospheric chemistry	<0.1
Ice nuclei (IN)	Has very special crystal structure	Causes freezing of supercooling water droplets	Natural dusts	>1
Small ions	Stable particle with an electric charge	Carries atmospheric electricity	All sources	>0.0015
Large particles	Special name			0.1–1
Giant particles	Special name			>1

hydrocarbons, and photochemical oxidants as ozone. In California, standards are also set for lead, hydrogen sulfide, and "visibility-reducing particles."

The main class of chemically identified carcinogens in the air is the polycyclic aromatic hydrocarbons (PAHs). Benzo-2-pyrene (BP) is the PAH which has been most widely studied, but it is not recognized as either a complete or conclusive indicator of the carcinogenic potential of pollution, since BP is not a carcinogen as such. It must be metabolically activated to a chemically reactive electrophilic species. An enzyme system believed involved in these changes is the aryl hydrocarbon hydroxylase (AHH) system.[10]

Eye irritation is one of the more conspicuous and obvious effects of the aerosols which constitute smog. Olefins are especially reactive when oxides of nitrogen react photochemically with the hydrocarbons to produce smog. Solvents play a major role in such formation, and have been rated for their tendency to produce a smog capable of eye irritation at 20-ppm and at 5-ppm concentrations. At the 20-ppm level, the following ranking of substances has been reported (in decreasing order of irritation):

Methyl isobutyl ketone
Trichloroethylene

Xylene

Methyl ethyl ketone

Hydrocarbon fractions (Six representative samples with boiling ranges from 110 to 264°F to 355 to 488°F constituted the hydrocarbon fractions in this research.

Methanol

Toluene

Hexane

Ethanol

Isopropanol

It is recognized that the composition of gases and aerosols in the air may be measured and monitored by well-established analytical methods, and hence a cause/effect relationship of relative safety or potential hazards, such as oxygen deficiency or carbon monoxide concentrations, can be performed relatively easily. On the other hand, most particulate or aerosol matter is measured by its total suspended particulate content which includes all filterable particles. Molecular and ionic species are often lumped together, precluding cause/effect studies where chemical composition is important. For this reason alone, one must examine carefully any analytical data on air for its implications as "breathing air."

Often overlooked is the fact that aerosols may travel thousands of miles and affect local conditions. For example, during September 1960 and October 1960, radioactive debris or "fallout" particles from open-air nuclear weapon testing in Nevada caused a highly significant increase in the radiation levels of the Northeast, including the tri-city area of Albany, Schenectady, and Troy, New York. Recently, the World Climate Conference in Geneva heard reports suggesting that smog and dust from industrial Europe and China may account for the haze that persists over Alaska, Greenland, and the Arctic Ocean every spring. "Acid rain" is another example.

TABLE 6.3. Particle Size Versus Nasal Penetration[a]

Particle Size (μ)	Nasal Penetration (%)
1.0	100
1.2	100
1.5	78
2.0	57
2.5	41
3.0	31
3.5	26
4.0	22
5.0	17

[a]EPA Proposes Range of Options to Revise Standards That Limit Particles in the Air, *Wall Street Journal*, p. 8, March 12, 1984. (Discusses proposed regulation options to reduce smaller particles, 10 microns or smaller.)

To appreciate the complexity of evaluating the actions and effects[12,13] of respiratory hazards, and protection from them, whether from air pollution or from industrial airborne contamination, a short description of the human respiratory system is in order. The breathing air normally enters the body through the nostrils. Larger particles or aerosols may be retained on the unciliated anterior portion of the nose, while other particles or aerosols, as well as the gases, pass through a web of nasal hairs and then flow through the narrow passages around the turbinates. The "inspired air" is warmed, moistened, and partially depleted of particles with aerodynamic diameters greater than 1 μm by sedimentation and impaction on nasal hairs and passages (see Fig. 6.1). The beating action of the cilia propels the inspired

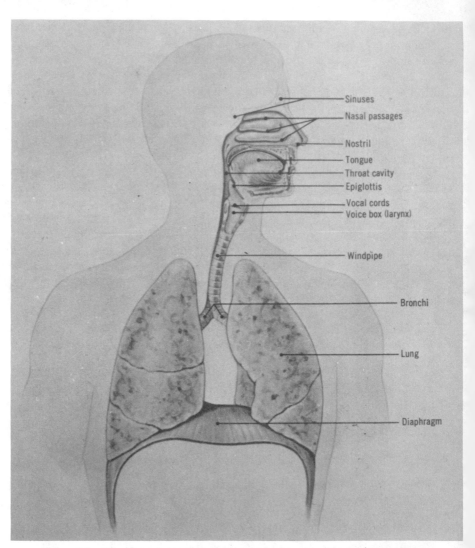

Figure 6.1. Human respiratory tree, showing main elements of the respiratory system.

air towards the pharynx. Deposited insoluble particles are transported by the mucus and soluble particles may dissolve in it.

Particles inhaled through the nose and deposited in the nasopharynx or particles inhaled through the mouth and deposited in the mouth and oropharynx are swallowed in minutes into the gastrointestinal tract.

The inspired air now proceeds down the trachea, or windpipe, which divides into two branches or primary bronchi leading in turn to upper and lower lobar bronchi for the right and left. The airway diameter decreases, but the number of tubes increase. The total cross section for flow increases and the air velocity decreases.

At the low velocities in the smaller airways, particles deposit by sedimentation and diffusion. Inert nonsoluble particles deposited on normal ciliated airways are cleaned within 1 day by transport on the moving mucus to the larynx. Soluble particles are cleaned much faster, presumably by bronchial blood flow.[14]

Gas exchange occurs in the acini of the lung parenchyma, or the portions of the lungs from the first order of respiratory bronchioles down to the alveoli. These respiratory bronchioles originate from the terminal bronchioles which are the smallest airways not concerned with gas exchange. The system of airways leading to the acini does not participate in the gas exchange and is called the "dead space." Inhaled particles may be deposited either in the lung parenchyma (the bronchioles, atrial sacs, and alveoli) or in the dead space. Smaller inhaled particles may be breathed in and out of the respiratory tract without deposition.[15] Examples of respiratory hazards and protective devices, by classes, which protect against them, are shown in Fig. 6.2.

6.1. DETECTION OF ATMOSPHERIC CONTAMINANTS

Odor is the simplest sense for detecting certain contaminants in the air, since *pure* air should smell clean or odorless. Odor is not a dependable criteria of air quality, however, since it is often misleading or unreliable. Certain substances such as acrolein, ammonia, bromine, chlorine, formaldehyde, acetic acid, acetic anhydride, sulfur dioxide and trioxide, and hydrogen chloride have distinctive odors at relatively low concentrations. Some gases and vapors, such as fluorine and hydrogen fluoride, are so immediately corrosive to the upper respiratory tract that odor is not a safe warning. Certain gases, such as arsine, phosphine, stibine, hydrogen cyanide, nitric oxide, and hydrogen sulfide may be encountered above permissible "safe" inhalation limits before a "fresh" nose can respond. The ability of several of these gases, including hydrogen sulfide, to cause olfactory fatigue is well recognized. The olfactory nerves are rapidly overcome, and can no longer sense changes in concentrations. A concentration of 1000-ppm hydrogen sulfide causes loss of consciousness in seconds from a few breaths (see Table 6.4, part d). This ability of hydrogen sulfide to rapidly overcome detection is doubtlessly a factor in the continuing fatal exposures which have occurred over the years from exposure to this gas.[16,17]

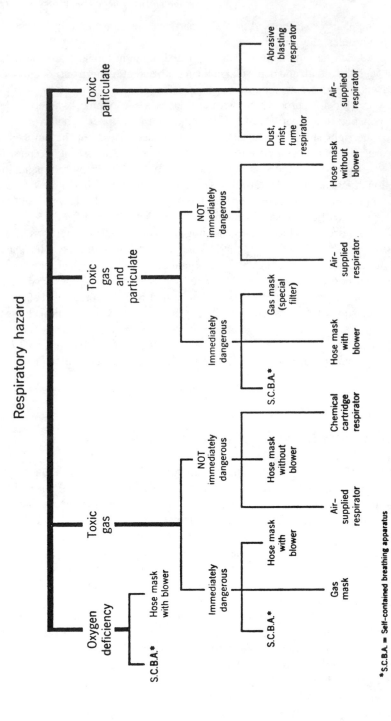

Figure 6.2. Respiratory hazards and the classes of protective devices available.

*S.C.B.A. = Self-contained breathing apparatus

112

At the other end of the scale of odors, carbon monoxide is, for all practical purposes, odorless and colorless. Inhalation can often result in significant, and even fatal, consequences (see Table 6.4, part a). The action of CO is due to its ability to compete with oxygen for hemoglobin-binding sites in the blood, and subsequent transfer to the body tissues. The degree of competition of oxygen versus CO is reflected in that CO has an affinity for the hemoglobin complex that is over 200 times greater than the oxygen–hemoglobin affinity. Once carboxyhemoglobin is formed, a considerable amount of time is required for the complex to dissociate. Until this happens, oxygen binding or transport is impossible. Thus the main effect of increased blood-CO levels is tissue hypoxia, or oxygen deficiency.

As the tissue demand for oxygen increases, increased stress on the heart is experienced because of an increased cardiac output. Therefore the heart is the first target organ during CO inhalation. This is especially significant for those individuals who may already be experiencing cardiovascular difficulty. The other organ system most rapidly affected by hypoxia is the central nervous system (CNS).

As the carboxyhemoglobin level increases, a progression of symptoms is noted. Nausea and headache symptoms start at blood-carboxyhemoglobin levels of 15%. At levels of 25%, changes in the electrocardiogram occur. At 40%, unconsciousness will usually ensue. Blood-carboxyhemoglobin levels of 66% or higher are usually considered fatal, although death has resulted from extended exposures at percentage levels in the 30s, such as during attempted escape from fire gases which may contain high percentages of CO.

After intense, high-level exposure to CO has ceased, several pathologic events can occur. The most serious is the development of cerebral edema, which can be life threatening if left untreated. This excessive accumulation of fluid in the brain is a result of increased permeability of the capillaries due to the change in oxygen tension. Within several days, confusion and other signs of mental deterioration can ensue. Serious mental deficits that are often irreversible may result from prolonged intense exposure. In addition, transient cardiac arrhythmias and enzyme elevations may also develop. The danger of asphyxiation in the unconscious person is increased by such serious complications as aspiration pneumonia and laryngeal edema.[18a-c]

While hydrogen sulfide and carbon monoxide are extremely serious problems, other gases are of concern. Sulfur dioxide (Table 6.4, part c), oxides of nitrogen, and ammonia (Table 6.4, part b) are also serious potential gases insofar as respiration is concerned. The first two are acidic in nature and the third is basic in nature. If an individual is occupationally exposed to these materials, the entire respiratory tree is affected, with reactions ranging from sneezing and coughing to severe bronchoconstriction and cessation of respiration. Because of this induced change in normal respiratory physiology, cardiovascular responses which may follow include increased blood pressure and pulse rate.

Acute, intense exposure to these gases may also produce pulmonary edema, an increase in intercellular and interstitial fluid. This edematous condition may result in a variety of physiologic malfunctions, including decreased lung compliance, hypoxemia, ventilation/perfusion mismatch, and respiratory alkalosis followed by metabolic acidosis.

TABLE 6.4. Symptoms of Exposures to Selected Gases[a]

Effect	Concentration (ppm)
a. Carbon Monoxide	
TLV-TWA/STEL	*50/400*
Slight headache in some cases	0–200
After 5–6 hr, mild headache, nausea, vertigo, and mental symptoms	200–400
After 4–5 hr, severe headache, muscular incoordination weakness, vomiting, and collapse	400–700
After 3–5 hr, severe headache, weakness, vomiting, and collapse	700–1000
After 1.5–3 hr, coma, and breathing still fairly good unless poisoning has been prolonged	1100–1600
After 1–1.5 hr, possibly death	1600–2000
After 2–15 min, death	5000–10,000
b. Ammonia	
TLV-TWA/STEL	*25/35*
Least detectable odor	53
Least amount causing immediate irritation to the eyes	698
Least amount causing immediate irritation to the throat	408
Least amount causing coughing	1720
Maximum concentration allowable for prolonged exposure	100
Maximum concentration allowable for short exposure ($\frac{1}{2}$–1 hr)	300–500
Dangerous for even short exposure ($\frac{1}{2}$ hr)	2500–4500
c. Sulfur Dioxide	
TLV-TWA/STEL	*2/5*
Least amount causing detectable odor	3–5
Least amount causing immediate eye irritation	20
Least amount causing immediate throat irritation	8–12
Least amount causing coughing	20
Maximum concentration allowable for prolonged exposure	10
Maximum concentration allowable for short (30 min) exposure	50–100
Amount dangerous for even short exposure	400–500
d. Hydrogen Sulfide	
TLV-TWA/STEL	*10/15*
Eye and respiratory tract irritation after exposure of 1 hr	50–100
Marked eye and respiratory tract irritation after exposure of 1 hr	200–300
Dizziness, headache, nausea, etc., within 15 min and loss of consciousness and possible death after 30–60 min exposure	500–700
Rapidly produces unconsciousness and death occurs a few minutes later	700–900
Death is apparently instantaneous	1000–2000

(Apparent death from H_2S is not irreversible as prompt and efficient artificial respiration may restore life.)

As continuous exposure to these irritant gases increases, definite clinical disease patterns may be seen, such as bronchiolitis, bronchitis, and pneumonia, as well as adverse changes in the teeth, eyes, and skin. An individual who is experiencing respiratory disease and who continues to be exposed to these irritant gases may ultimately experience irreversible respiratory pathology. (Table 6.4 parts b–d, presents a summary of some of the inhalation data regarding these gases.)

The effect of exposure to gases is enhanced by factors such as heavy labor, high environmental temperature, and increased altitude (over 2000 ft). Susceptibility is greatest in the very young, the elderly, those with cardiac or chronic respiratory disease, and pregnant women.

The odor intensity and physiological response produced by paraffin hydrocarbons have been studied in terms of the potential warning of possible concentration build-up leading to an explosive mixture with air. The odors of heptane and hexane are easily noticed in concentrations below their lower flammable limits. Heptane and hexane vapors produce distinct symptoms. The odor of pentane is indistinct, and that of ethane and propane practically absent in lower flammable limit mixtures.

Historically, animals such as canaries and Japanese waltzing mice have been used as warnings for "bad air" in mines and other confined operations. The Davy mine safety lamp, an indicator of oxygen deficiency or methane atmospheres, has been largely displaced by portable as well as fixed-station instrumentation to detect and record low or changing concentrations of gases, vapors, dusts, and aerosols. The availability of gas detector tubes for use with personal samplers worn by workers has been a major advance. In addition, personal gaseous "film badges" have been developed and are in use for monitoring exposures to many gases and vapors.

The "Odor Threshold Manual" of the American Standards for Testing and Materials is an excellent compilation of published data on odors.[19] It must always be recognized that some persons have virtually no ability to detect an odor, while others have a high sensitivity which may produce serious physiological as well as emotional effects.[20]

Sources. For carbon monoxide, ammonia, and sulfur dioxide: Kirk-Othmer, *Encyclopedia of Chemical Technology*, 2nd ed., Vol. 2, p. 291; Vol. 4, p. 442; Vol. 19, p. 417, Wiley, New York, 1980. For hydrogen sulfide: *International Oil Tanker and Terminal Safety Guide*, The Institute of Petroleum, London, England, 1980. See also Lawrence, R. W. and Long, S. E., Guidelines for Installing Sensors For Monitoring Hazardous Gases, *Plant Engineering*, Vol. 38, No. 6, pp. 188–191, March 8, 1984.

[a]Instrumentation for remote monitoring of toxic gases (such as the 50-ppm H_2S and 150-ppm CO range), as well as for flammable vapors and gases in air are now available from a number of supplies, including Control Instruments, 25 Law Drive, Fairfield, NJ 07006. Instrument suppliers should be contacted for details of their devices.

[b]TLV Values are from the 1983–1984 list of the American Conference of Governmental Industrial Hygienists. TWA = time-weighted average and STEL = short-term exposure limit.

6.2. PROTECTION

To prevent or reduce exposures to airborne substances, the substances should be confined to closed systems, or removed from the breathing zone by properly designed vents or fume hoods. They should then be moved to outside the building away from air intakes and other inhabited areas, after proper dilution or filtration or scrubbing to reduce the level of discharged effluent at the stack to an acceptable level. The use of filters and other pollution-control devices should be carefully engineered.[21] To illustrate, discharge to the roof level may be satisfactory most of the time, but if the wind-directional rose indicates that, even for a small percentage of time, the effluent air may be recycled or reach a populated area, serious concern should be given to the air movements. In one specific operational problem, ethylene dibromide was being exhausted from an animal inhalation laboratory at a concentration of 20-ppm. Since it was near a populated area, it was desired to reduce the stack concentration to 1-ppm. Activated charcoal bed filters were installed into the vent system at a cost of $10,000, part of which was for changes made to the duct system to accommodate the filters. The filters performed successfully, and frequent monitoring was instituted to ensure that saturation or breakthrough did not occur between the scheduled monthly filter replacements.

If confinement or removal by adequate properly engineered and operated ventilation systems is not possible, personal protective equipment on a temporary basis should be considered. We stress the word "temporary" since respiratory protection can seldom be relied on for long periods of time in hazardous exposures, unless highly unusual control procedures are established and rigorously enforced. Where the risks are known to be high, as in the possibility of breathing plutonium or dioxin, or with aerosols containing carcinogens, programs of respiratory protection have been successfully instituted. But in areas or situations where less obvious hazards exist, respiratory protection is successful only if well-planned and implemented programs of education, maintenance, and enforcement are in use.[22,23]

6.3. AIR SUPPLY

Where air-supplied respiratory protection is used, the very important problem of the purity of the breathing air must be considered. In the past, fatal accidents have occurred when an air-supplied respirator was plugged into a plant air supply, which had, without warning, been changed to a nitrogen supply while work was being done on the compressor. In other cases, cylinders of breathing air were, in fact, filled with nitrogen, carbon dioxide, and butane. In some instances, air has been produced by mixing compressed oxygen with compressed nitrogen, but all such mixtures should be suspect until it can be shown that they do, in fact, contain a mixture which is within the recognized limits for breathing purposes. While oxygen deficiency is the more obvious potential error, the problems associated with oxygen above 25%, from a fire hazard viewpoint, should not be overlooked. Air-supplied

Figure 6.3. Air-supplied hood. Note air connections at waist level on left. Quality of the supplied air is an essential aspect of such devices. Courtesy of MSA.

hoods are frequently used for short-duration tasks. A typical example is shown in Fig. 6.3.

Legal standards exist for air supply from a compressor: OSHA specification 1910.94 (6) notes that the air for abrasive-blasting respirators shall be free of harmful quantities of dusts, mists, or noxious gases, and shall meet the requirements for air purity set forth in ANSI Z9.2-1960. The air from the regular compressed air line of the plant may be used for the abrasive-blasting respirator if: (1) a trap and carbon filter are installed and regularly maintained to remove oil, water, scale, and odor; (2) a pressure-reducing diaphragm or valve is installed to reduce the pressure to the requirements of the particular type of abrasive-blasting respirator; and (3) an automatic control is provided to either sound an alarm or shut down the compressor in case of overheating. In the OSHA specification 1910.134, the quality of air is specified. Oxygen shall meet the requirements of the U.S. Pharmacopaeia for medical or breathing oxygen.[24] The breathing air shall meet at least the requirements of the specification for the Grade D breathing air as described in Compressed Gas Association (CGA) Commodity Specification G-7.1-1973. Compressed oxygen shall not be used in supplied-air respirators or in open-circuit self-contained breathing apparatus (SCBA) that have previously used compressed air. Oxygen must never be used with air-line hoods or respirators.[25] If a compressor is used, it shall be of breathing-air type, constructed and situated so as to avoid entry of contaminated air into the system and suitable in-line air purifying absorbent beds and filters installed to further assure the breathing air quality. If an oil-lubricated compressor is used, it shall have a high-temperature or carbon monoxide alarm, or both. If only a high-temperature alarm is used, the air from the compressor shall be frequently tested for carbon monoxide to ensure that it meets the specifications.[24] Where air is used for diving, the specifications of the U.S. Navy Diving Manual, *NAV-SHIPS 250-538*, should be followed. The CGA and Navy Standards are summarized below:

| | Maximum Allowable Level | |
Contaminant	CGA's Grade D	U.S. Navy
Carbon monoxide (ppm)	20	20
Carbon dioxide (ppm)	1000	500
Oil vapor (mg/m³)	5	5

The states of California, New Hampshire, New Jersey, New York, and Washington have even lower limits, and specify that the SCBA air be free from odors and other contaminants.

Classification and Description of Respiratory Protective Devices

Based on the mode of operation, several types of respiratory protective devices are available. Table 6.5 describes the general classification and the mode of operation of these respirators.

TABLE 6.5.   Classification and Description of Respirators by Mode of Operation

Atmosphere-Supplying Respirators

A respirable atmosphere independent of the ambient air is supplied to the wearer.

Self-Contained Breathing Apparatus (SCBA)

Supply of air, oxygen, or oxygen-generating material is carried by the wearer. Normally it is equipped with full facepiece, but may be equipped with a quarter-mask facepiece, half-mask facepiece, helmet, hood, or mouthpiece and nose clamp.

1.   Closed-Circuit SCBA (oxygen only)

    a.   *Compressed- or Liquid-Oxygen Type Equipped With a Facepiece or Mouthpiece and Nose Clamp.*   High-pressure oxygen from a gas cylinder passes through a high-pressure reducing valve and, in some designs, through a low-pressure admission valve to a breathing bag or container. Liquid oxygen is converted to a low-pressure

TABLE 6.5. *(Continued)*

Atmosphere-Supplying Respirators

gaseous oxygen and delivered to the breathing bag. The wearer inhales from the bag through a corrugated tube connected to a mouthpiece or facepiece and a one-way check valve. Exhaled air passes through another check valve and tube into a container of a carbon-dioxide-removing chemical and reenters the breathing bag. Make-up oxygen enters the bag continuously or as the bag deflates sufficiently to actuate an admission valve. A pressure relief system is provided and a manual bypass system and saliva trap may be provided depending upon the design.

b. *Oxygen-Generating Type.* Equipped with a facepiece or mouthpiece and nose clamp. Water vapor in the exhaled breath reacts with chemical in the canister to release oxygen to the breathing bag. The wearer inhales from the bag through a corrugated tube and one-way check valve at the facepiece. Exhaled air passes through a second check-valve breathing-tube assembly into the canister. The oxygen-release rate is governed by the volume of exhaled air. Carbon dioxide in the exhaled breath is removed by the canister fill.

2. Open-Circuit SCBA (compressed air, compressed oxygen, liquid air, liquid oxygen). A bypass system is provided in case of regulator failure except on escape-type units.

a. *Demand Type.*[a] Equipped with a facepiece or mouthpiece and nose clamp. The demand valve permits oxygen or airflow only during inhalation. Exhaled breath passes to ambient atmosphere through a valve(s) in the facepiece.

b. *Pressure-Demand Type.*[b] Equipped with facepiece only. Positive pressure is maintained in the facepiece. The wearer may have the option of selecting the demand or pressure-demand mode of operation.

Supplied-Air Respirators

1. Hose mask is equipped with a respiratory-inlet covering (facepiece, helmet, hood, or suit), nonkinking breathing tube, rugged safety harness, and a large-diameter heavy-duty nonkinking air-supply hose. The breathing tube and hose are securely attached to the harness. A facepiece is equipped with an exhalation valve. The harness has provision for attaching a safety line.

a. *Hose Mask With Blower.* Air is supplied by a motor-driven or hand-operated blower. The wearer can continue to inhale through the hose if the blower fails. Up to 300 ft (91 m) of hose length is permissible.

b. *Hose Mask Without Blower.* The wearer provides motivating force to pull air through the hose. The hose inlet is anchored and fitted with a funnel or like object covered with a fine mesh screen to prevent entrance of coarse particulate matter. Up to 75 ft (23 m) of hose length is permissible.

2. Air-line respirator provides respirable air through a small-diameter hose from a compressor or compressed-air cylinder(s). The hose is attached to the wearer by a belt and can be detached rapidly in an emergency. A flow control valve or orifice is provided to govern the rate of airflow to the wearer. Exhaled air passes to the ambient atmosphere through a valve(s) or opening(s) in the enclosure (facepiece, helmet, hood, or suit). Up to 300 ft (91 m) of hose length is permissible.

a. *Continuous-Flow Class.* Equipped with a facepiece, hood, helmet, or suit. At least 115 liters (4 ft^3) of air per minute to tightfitting facepieces and 170 liters (6 ft^3) of air per minute to loosefitting helmets, hoods, and suits are required. Air is supplied to a suit through a system of internal tubes to the head, trunk, and extremities through valves located in appropriate parts of the suit.

TABLE 6.5.   (*Continued*)

Atmosphere-Supplying Respirators

b. *Demand Type.*[a]   Equipped with a facepiece only. The demand valve permits flow of air only during inhalation.

c. *Pressure-Demand Type.*[b]   Equipped with a facepiece only. A positive pressure is maintained in the facepiece.

Combination Air-Line Respirators With an Auxiliary Self-Contained Air Supply

Includes an air-line respirator with an auxiliary self-contained air supply. To escape from a hazardous atmosphere in the event the primary air supply fails to operate, the wearer switches to the auxiliary self-contained air supply. Devices approved for both entry into and escape from dangerous atmospheres have a low-pressure warning alarm and a self-contained air supply.

Air-Purifying Respirators

Ambient air prior to being inhaled is passed through a filter, cartridge, or canister which removes particles, vapors, gases, or a combination of these contaminants. The breathing action of the wearer operates the nonpowered type of air-purifying respirator. The powered type contains a blower which is stationary or carried by wearer. It passes ambient air through an air-purifying component and then supplies purified air to the respiratory-inlet covering. The nonpowered type is equipped with a facepiece or mouthpiece and nose clamp. The powered type is equipped with a facepiece, helmet, hood, or suit.

Vapor- and Gas-Removing Respirators

Equipped with cartridge(s) or canister(s) to remove a single vapor or gas (e.g., chlorine gas), a single class of vapors or gases (e.g., organic vapors), or a combination of two or more classes of vapors or gases (e.g., organic vapors and acid gases) from the air.

Particulate-Removing Respirators

Equipped with filter(s) to remove a single type of particulate matter (e.g., dust) or a combination of two or more types of particulate matter (e.g., dust and fume) from air. Filter may be a replaceable part or a permanent part of the respirator. Filter may be single-use type or reusable type.

Combination Particulate-, Vapor-, and Gas-Removing Respirators

Equipped with cartridge(s) or canister(s) to remove particulate matter, vapors, and gases from the air. The filter may be a permanent part or a replaceable part of a cartridge or canister.

Combination Atmosphere-Supplying and Air-Purifying Respirators

An atmosphere-supplying respirator with an auxiliary air-purifying attachment provides protection in the event the air supply fails. An air-purifying respirator with an auxiliary self-contained air supply is used in case the atmosphere unexpectedly exceeds safe conditions for use of an air-purifying respirator.

[a]Equipped with a demand valve that is activated on initiation of inhalation and permits the flow of breathing atmosphere to the facepiece. On exhalation, pressure in the facepiece becomes positive and the demand valve is deactivated.

[b]A positive pressure is maintained in the facepiece by a spring-loaded or balanced regulator and exhalation valve.

The capabilities and limitations of the various types of respirators are listed in Table 6.6 and are discussed in detail below.

Self-Contained Breathing Apparatus (SCBA)

The self-contained breathing apparatus is generally understood to furnish the maximum protection against respiratory hazards. It is designed to supply complete respiratory protection (except for gases or vapors which may penetrate the skin). When used in atmospheres where gases or vapors with significant action of toxicity through the skin is involved, such as hydrogen cyanide, the respiratory protective device must be supplemented by complete skin protection of an impervious type, in addition to respiratory protection. Until a fully recognized and agreed-upon standard for testing impervious clothing is implemented, suspicion of the effectiveness of protection should always be raised, especially when details of what is involved and the concentration involved may be lacking (see Chap. 7).

Since the SCBA has no external connection to an air supply, it is obviously the device of choice for emergency situations and entry into confined spaces (see Figs.

TABLE 6.6.   Capabilities and Limitations of Respirators

Atmosphere-Supplying Respirators

Atmosphere-supplying respirators provide protection against oxygen deficiency and toxic atmospheres. The breathing atmosphere is independent of ambient atmospheric conditions.

General Limitations: Except for some air-line suits, no protection is provided against skin irritation by materials such as ammonia and hydrogen chloride, or against absorption of materials such as hydrogen cyanide, tritium, or organic phosphate pesticides through the skin. Facepieces present special problems to individuals required to wear prescription lenses. Use of atmosphere-supplying respirators in atmospheres immediately dangerous to life or health is limited to specific devices under specified conditions.

Self-Contained Breathing Apparatus (SCBA)

The wearer carries his or her own breathing atmosphere.

Limitations: The period over which the device will provide protection is limited by the amount of air or oxygen in the apparatus, the ambient atmospheric pressure (service life is cut in half by a doubling of the atmospheric pressure), and work. Some SCBA devices have a short service life (less than 15 min) and are suitable only for escape (self-rescue) from an irrespirable atmosphere. Chief limitations of the SCBA devices are their weight or bulk or both, limited service life, and the training required for their maintenance and safe use.

1. *Closed-Circuit SCBA.*   The closed-circuit operation conserves oxygen and permits longer service life at reduced weight. A negative pressure in the respiratory-inlet covering is created during inhalation in most closed-circuit devices and may permit inward leakage of contaminants.
2. *Open-Circuit SCBA.*   Demand and pressure-demand. The demand type produces a negative pressure in the respiratory-inlet covering during inhalation, whereas the pressure-demand type maintains a positive pressure in the respiratory-inlet covering during inhalation and is less apt to permit inward leakage of contaminants.

TABLE 6.6. *(Continued)*

Atmosphere-Supplying Respirators

Supplied-Air Respirators

The respirable-air supply is not limited to the quantity the individual can carry and the devices are lightweight and simple.

Limitations: Limited to use in atmospheres from which the wearer can escape unharmed without aid of the respirator. The wearer is restricted in movement by the hose and must return to a respirable atmosphere by retracing route of entry. The hose is subject to being severed or pinched off.

1. *Hose Mask.* The hose inlet or blower must be located and secured in a respirable atmosphere.
   a. *Hose Mask With Blower.* If the blower fails, the unit still provides protection, although a negative pressure exists in the facepiece during inhalation.
   b. *Hose Mask Without Blower.* Maximum hose length may restrict application of device.

2. *Air-Line Respirators (Continuous-Flow, Demand, and Pressure-Demand Types).* The demand type produces a negative pressure in the facepiece on inhalation, whereas continuous-flow and pressure-demand types maintain a positive pressure in the respiratory-inlet covering and are less apt to permit inward leakage of contaminants. Air-line suits may protect against atmospheres that affect the skin or mucous membranes or that may be absorbed through the unbroken skin.

Limitations: Air-line respirators provide no protection if the air supply fails. Some contaminants, such as tritium, may penetrate the material of an air-line suit and limit its effectiveness. Other contaminants, such as fluorine, may react chemically with the material of an air-line suit and damage it.

Combination Air-line Respirator With an Auxiliary Self-Contained Air Supply

The auxiliary self-contained air supply on this type of device allows the wearer to escape from a dangerous atmosphere. This device with an auxiliary self-contained air supply is approved for escape and may be used for entry when it contains at least a 15-min auxiliary self-contained air supply.

Air-Purifying Respirators

General Limitations: Air-purifying respirators do not protect against oxygen-deficient atmospheres nor against skin irritations by, or sorption through, the skin of airborne contaminants. The maximum contaminant concentration against which an air-purifying respirator will protect is determined by the design efficiency and capacity of the cartridge, canister, or filter and the facepiece-to-face seal on the user. For gases and vapors, the maximum concentration for which the air-purifying element is designed is specified by the manufacturer, or is listed on labels of cartridges and canisters.

Nonpowered air-purifying respirators will not provide the maximum design protection specified unless the facepiece or mouthpiece/nose clamp is carefully fitted to the wearer's face to prevent inward leakage. The time period over which protection is provided is dependent on the canister, cartridge, or filter type; concentration of contaminant; humidity levels in the ambient atmosphere; and the wearer's respiratory rate.

TABLE 6.6. *(Continued)*

Air-Purifying Respirators

The proper type of canister, cartridge, or filter must be selected for the particular atmosphere and conditions. Nonpowered, air-purifying respirators may cause discomfort due to a noticeable resistance to inhalation. This problem is minimized in powered respirators. Respirator facepieces present special problems to individuals required to wear prescription lenses. These devices do have the advantage of being small, light, and simple in operation.

Use of air-purifying respirators in atmospheres immediately dangerous to life or health is limited to specific devices under specified conditions.

Vapor- and Gas-Removing Respirators

A small, lightweight device that can be donned quickly.

Limitations: No protection is provided against particulate contaminants. A rise in canister or cartridge temperature indicates that a gas or vapor is being removed from the inspired air. An uncomfortably high temperature indicates a high concentration of gas or vapor and requires an immediate return to fresh air. Should avoid use in atmospheres where the contaminant(s) lacks sufficient warning properties (i.e., odor, taste, or irritation).

1. *Full-Facepiece Respirator.* Provides protection against eye irritation in addition to respiratory protection.
2. *Half- and Quarter-Mask Respirator.* Not for use in atmospheres immediately dangerous to life or health unless it is a powered-type respirator with escape provisions. A fabric covering (facelet) available from some manufacturers shall not be used.
3. *Mouthpiece Respirator.* Shall be used only for escape applications. Mouth breathing prevents detection of contaminant by odor. Nose clamp must be securely in place to prevent nasal breathing.

Particulate-Removing Respirators

A small, lightweight device that can be donned quickly.

Limitations: Protection against nonvolatile particles only. No protection against gases and vapors.

1. *Full-Facepiece Respirator.* Provides protection against eye irritation in addition to respiratory protection.
2. *Half- and Quarter-Mask Respirator.* Not for use in atmospheres immediately dangerous to life or health unless it is a powered-type respirator with escape provisions.
3. *Mouthpiece Respirator.* Shall be used only for escape applications. Mouth breathing prevents detection of contaminant by odor. Nose clamp must be securely in place to prevent nasal breathing.

Combination Particulate-, Vapor-, and Gas-Removing Respirators

The advantages and disadvantages of the component sections of the combination respirator as described above apply.

Combination Atmosphere-Supplying and Air-Purifying Respirators

The advantages and disadvantages, expressed above, of the mode of operation being used will govern. The mode with the greater limitations (air-purifying mode) will mainly determine the overall capabilities and limitations of the respirator since the wearer may for some reason fail to change the mode of operation even though conditions would require such a change.

Source. "Standards for Respiratory Protection," ANSI Z88.2-1980.

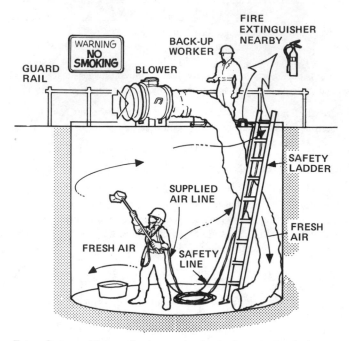

Figure 6.4.   Entry of personnel into confined spaces (such as tanks, sewers and other areas of restricted air movement) should include a SCBA and a back-up person also wearing a SCBA for possible rescue. Courtesy of Plant Engineering.

6.4 and 6.5). Even when wearing a SCBA, asphyxiation is possible if the facepiece is not properly fitted, as from facial hair or improper orientation. The question of endangering another person should never be overlooked. Gasoline vapors were the hazardous atmosphere described by F. H. Halvorsen, "Death by Asphyxiation in Tank Vessel," pp. 199–201 *Proceedings of the Marine Safety Council* **40,** No. 8 (September 1983) (U.S. Coast Guard, Washington, D.C. 20593).

The first self-contained breathing apparatus to be approved by the U.S. Bureau of Mines in 1916 was the 2-hr rebreathing type. It was intended for mine-rescue work, and is still used today for that purpose. Originally, a mouth-breathing facepiece with a nose clip to seal the nose was used, but in more recent times a full facepiece for these devices has been approved. When the NIOSH was formed in 1971, the approval system for these devices which had been the U.S. Bureau of Mines (and the Department of Agriculture) was gradually moved to NIOSH, and the certification of devices became an activity at NIOSH's laboratory in Morgantown, West Virginia.

The most widely used SCBA today is a demand open-circuit type, consisting of an air cylinder, an air line and regulator, and facepiece (see Fig. 6.6). Depending on the size and pressure in the cylinder, the supply may last from 5 to 30 min. In recent times, it has been recognized that the demand principle [i.e., air was available

Figure 6.5. Self-contained breathing apparatus (SCBA) escape tank used with air-supplied mask and safety line. Courtesy of Scott.

only when the wearer inhaled, and was cut off when the wearer exhaled (often creating a negative pressure in the facepiece which could encourage leakage from the hazardous external atmosphere)] is less desirable than a positive-pressure device, which has a continuous flow of air into the breathing zone to ensure against negative pressure at any time. The positive-pressure device has now been recommended for the fire services and any other exposure where extreme care must be taken to prevent toxic gases from entering the facepiece, such as work on a survey of toxic-waste sites. SCBA devices are usually of 30-min duration, but smaller devices with air supplies of 5, 10, 15, and 20 min have been available at various times, and may have specific application for quick entry and inspection purposes.[25] The impact of the NASA space program development on the SCBA should be noted, since the aluminum cylinder, wound with fiberglass reinforcement, is significantly lighter than the traditional steel cylinder, and is coming into wide use. The components of a typical SCBA are clearly shown in Fig. 6.7.

Figure 6.6.   Self-contained breathing apparatus (SCBA) with cylinder worn on back, providing 30 to 45 minutes air supply. Such devices may be either demand or constant-flow type. Aluminum fiberglass-wound reinforced cylinders are now available which are lighter in weight. Courtesy of Guardian.

Closed-circuit SCBA have evolved and developed from the mine-rescue models, and are now available with up to a 4-hr service life. Closed circuit equipment obtains this long life from the oxygen supply of a liquid, gaseous, or chemical solid. During World War II, an oxygen-breathing apparatus (OBA) was developed, using potassium superoxide as the oxygen source, which also adsorbed carbon dioxide.[26] These devices were further developed, and were rated at times of 30, 45, and 60 min (see Fig. 6.8). A chlorate starting "candle" was added to aid in starting, especially in very cold weather.[27] In this writer's opinion, the devices using the superoxide demand much more careful training, especially in cold climates. The very real explosion potential of a carelessly discarded canister in contact with oil or other combustibles must not be overlooked. *Once a canister has been used, even for a short period, it should be promptly and carefully destroyed, and not used for practice or training. The disposal must follow exactly the instructions, if a fire or explosion is to be avoided.*

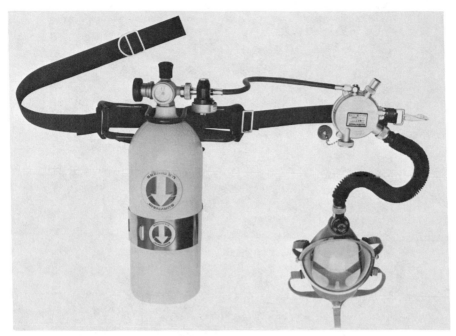

Figure 6.7.   Components of the SCBA. Courtesy of U.S. Divers.

Supplied-Air Respirators

If a limited area is to be entered, and a hose or other connection is not a problem, air-supplied respirators or hoods have real advantages. The hose mask, which has been used for years for certain operations where long-time work is important consists of a large-diameter hose, anchored in uncontaminated air and connected to the facepiece. If the diameter is sufficiently large, and the work requirements not excessive, the wearer experiences little resistance to breathing. For some applications, a positive blower may be included, either manually operated by another person (who also serves as a back-up or buddy) or by an electric-powered fan or blower, using power from a completely dependable source. The maximum flow of approved air masks is 150 liters/min. This flow will not maintain a positive pressure within the facepiece during periods of exertion by the wearer. Powered hose masks should be considered negative-pressure devices.

Air-line respirators deliver breathing air to the facepiece from a source which supplies pressure. Normally, a compressor is used to supply the air, but systems may be engineered by which cylinders of breathing air may be used (see Fig. 6.9). These cylinders may be demand, pressure-demand, or continuous-flow types. The demand device has an airflow regulator between the facepiece and the supply, so inhalation negative pressure activates the diaphragm, which opens the air supply. The negative pressure required to activate the air supply may be as much as 50

Figure 6.8.   Self-contained breathing apparatus using a solid oxygen source (similar to OBA). Starting "candle" is below the black circle in the center of the assembly. Used canisters must be disposed of properly to prevent difficulties (see text). Courtesy of General Electric Research Laboratory.

mm of water, and unless a constant-flow type is used, the possibility of facepiece leakage may be real. The facepieces may be quarter-, half-, or full facepieces. The NIOSH certification requires that a minimum of 115 liters/min be delivered to the facepiece on demand. One objection to air-line respirators, in addition to the limiting area, is the noise of the air passing into the facepiece.

Combination Self-Contained and Supplied-Air Respirators

For situations where prolonged times are necessary, and back-up escape essential, the SCBA may be supplemented with an air-line respirator, which can be plugged in or unplugged rapidly, and which serves as the air supply when engaged, thus conserving the breathing air in the cylinder (see Fig. 6.9.).

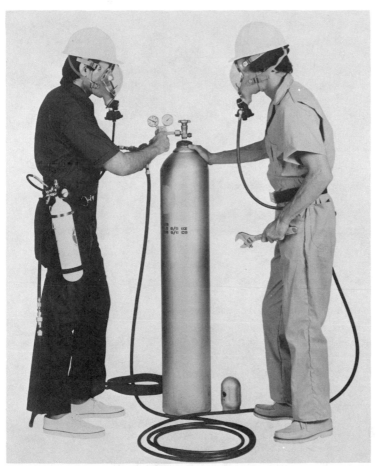

Figure 6.9. Air-line respirators may be supplied from cylinder "breathing air" or from compressor of proper type. Note wearer on left has cylinder for entry or escape. Courtesy of U.S. Divers.

Air-Purifying Respirators

If the air to be encountered is known to be within the limits of the protection factor (see Section 6.3) for air-purifying devices and has an adequate oxygen content (over 19%), an air-purifying respirator may be used. It must be noted that the unit will protect only against the specific substance or combinations in certain concentrations for which it was designed (e.g., aerosol filter-type respirators will afford no protection against gases and vapors), and that the canister or filter mask must be maintained in proper condition attached to a properly maintained facepiece which fits snugly over the area of contact to preclude leakage of the contaminated air (see Fig. 6.10). In addition, the very serious question can arise of one person using

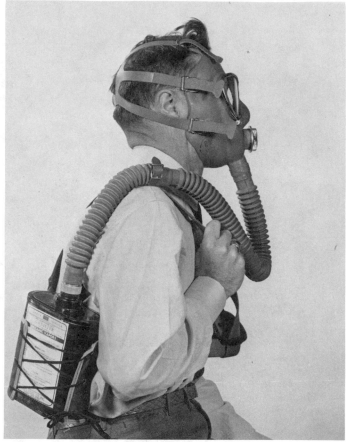

Figure 6.10.   Air-purifying "gas" mask for organic vapors with canister worn on back. This arrangement is especially effective for use where the face of the wearer must be in close proximity to gases or vapors, as in tank rodding or in opening drums and pipes. Courtesy of General Electric Research Laboratory.

another person's respirator, of inadequate instruction, of vague records of canister use, and of the unknown endpoint as to when a material has begun to penetrate the canister. In one specific case which occurred during an inspection of a carcinogen bioassay laboratory in which respiratory protection was required for entrance to the area, a visitor was furnished a respirator of the chemical cartridge type which had been used by other employees. During the inspection, which required 3 hr, the visitor was exposed to unknown (but supposedly small) concentrations of numerous suspect carcinogens, including ethylene chlorohydrin. No data were available as to how rapidly ethylene chlorohydrin penetrates a chemical cartridge respirator. It may or may not have been coincidence that the visitor suffered acute pulmonary edema that same evening, resulting in congestive heart failure and 7 months of disability.

Unless very careful control is implemented in a respirator program, such incidents can occur.

Aerosol filter-type respirators also require frequent changing of filters if breathing resistance becomes uncomfortably high when worn in areas containing excessive dust, mists, or fumes. Wearer acceptance will be a problem unless great care is taken to instruct, properly fit, and maintain these devices. Recently the OSHA has ruled that respirators need not be worn continuously during exposures to cotton dusts, but only to the extent necessary to reduce the total daily exposure level to the limits accepted. This rule was a direct result of the protest of workers who felt the respirators were too uncomfortable to wear for the whole work period.

Chemical cartridge respirators will safely protect only against the specific gases and vapors for which they were certified in nonemergency situations. Exposures to extremely toxic materials, such as acrolein, acrylonitrile, aniline, dimethylaniline, arsine, diborane, hydrogen cyanide, methyl fluorosulfate, carbon disulfide, phosphine, and other gases and vapors with high toxicity and low-warning odor levels or unknown characteristics should not be assumed safe when a chemical cartridge respirator is worn. Chemical cartridge respirators should not be used against gases and vapors which are odorless, or whose odor threshold is high, since odor is the only inplace warning of failure of the canister, of facepiece leakage, or of concentrations which are above the protection-factor limits of the respirator. Methyl chloride is an example of a gas whose warning properties are inadequate for practical warning purposes. Substances which are highly irritating to the eyes, such as ammonia and sulfur dioxide, as noted in Tables 6.4, parts b and c, require eye protection, such as a full facepiece, gas-tight goggles, or an sir-supplied hood. Several lacrimatory (tear-producing) substances, such as benzyl chloride, are in the same classification, and a respirator alone is clearly inadequate protection. Carbon monoxide cannot be stopped by a chemical cartridge, except by the Type-N universal mask (a seriously misleading term) and the miner's self-rescuer unit, both of which contain Hopcalite, a mixture of oxides of manganese, copper, cobalt, and silver. This mixture catalyzes the oxidation of carbon monoxide to carbon dioxide, provided that the mixture is properly activated and dry, and the carbon monoxide concentration is not excessive. These qualifications have lead to serious problems in the past, and for fire fighting and other services where *unknown* concentrations of carbon monoxide and other substances may be present, such as in toxic wastesites, *the Type-N mask is no longer recommended.*

Combination cartridges in masks and in respirators have the advantage that they provide some protection against more than one class of hazard, but the serious disadvantage of shorter service life for any particular hazard makes them less desirable.

Industrial canister-type gas masks, with canisters designed for specific substances or specific combinations, have the same limitations in general as chemical cartridge respirators. However, these gas masks are effective in concentrations of any specific gas or vapor, or members of the group of gases or vapors for which they were designed, of not more than 2% concentration in air or a 2% total concentration for a mixture of gases and vapors for which the canister is designed. The industrial-type ammonia canister mask is certified for a higher concentration.

Depending on the size of the canister, and the service for which it was designed, the service time varies. Protection against a combination of various gases and vapors, such as acid gases, organic vapors, and ammonia, can all be obtained in one canister, but the service life of such a combination is shorter than when used with one substance, that is, the life of an equivalent canister designed for the specific substance alone.

The only warning that a canister or cartridge is "exhausted" or spent is usually the sensory detection of the vapor or gas passing through the canister or cartridge to the wearer's nose. For carbon monoxide, which is odorless, canisters may include a timer or color indicator window. The self-rescuer, using the Hopcalite, does not have this warning feature. The only tangible indication that high concentrations are being encountered is the temperature increase, since the action of the CO on the

Figure 6.11. Prescription lenses attached to the post inside mask facepiece. Courtesy of General Electric Research Laboratory.

catalyst is highly exothermic. *Unless we have complete knowledge of the actual concentrations of what material are encountered, there may be a serious false sense of security in using canister-type or cartridge-type masks and respirators.*

Persons who need corrective lenses in order to work properly when wearing respiratory protection, especially full facepieces, should consider incorporation of the corrective lenses inside a facepiece in such a manner that the facepiece seal is not compromised. One method consists of wire frames which fit around the circumference of the facepiece sight area and hold 50-mm round lenses. Another approach suspends wire-frame goggles with short bows between holders molded into the facepiece. A third approach is a center post built inside the facepiece on which can be attached 40-mm rimless glasses. Such arrangements seldom encourage perfect alignment of the prescription lenses to the viewer's eyes, but, with adjustment, provide a sufficiently accurate fitting for most persons (see Fig. 6.11). Contact lenses offer a possible solution, but the hazards of contact lenses in increasing or aggravating a chemical burn in the eye, especially if the lenses are not immediately removed before irrigation with water, must be recognized. *It is the writer's belief that persons with serious limitation to their vision without lenses should not rely on respiratory devices which compromise their sight.*

Protection Factor

The protection factor (PF) is a relatively new concept which has introduced new understanding to the respiratory field. The development group at Los Alamos became concerned about this problem as related to protection from highly toxic aerosols, such as plutonium dust. Edwin C. Hyatt has published the basis for the PF.[28] The formula for calculation of the PF is

$$PF = \frac{\text{ambient air concentration}}{\text{concentration inside facepiece or enclosure}}$$

The PF concept is based largely on the premise that one must know how efficient the device is in actual use, and how much protection it really delivers. Over the years it has been observed that many wearers of respirators deliberately slip the straps so the face seal is loose and permits leakage, or they may even entirely negate the device as suggested by Fig. 6.12. For these reasons, we emphasize the importance of the fit test to ensure that the wearer has an adequate seal all around the face, and that the device, whether it is air-purifying or air-supplied, is functioning as it was designed. The original American designs, facial measurements, and relationships of facial features were based on American male anthropometric panels in World War I, when military gas masks were being developed after the first use of chlorine and phosgene at Ypres, Belgium in 1916. The proper fitting of the current wearers of respiratory protective devices has been made more acute by the increasingly high percentage (41%) of women workers, as well as the influx of workers from Asia whose measurements and proportions differ significantly from the American male norms. The fact is that many facepieces have not fitted properly, and any effective respiratory program must assign a high priority to fitting. Hyatt

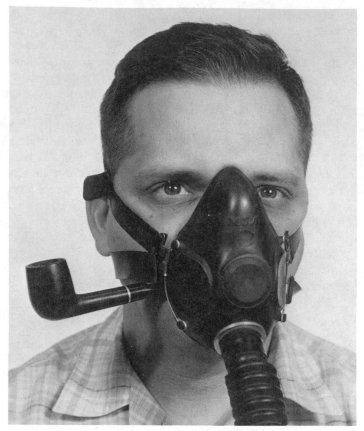

Figure 6.12. Respiratory protection can be easily negated by improper fit of facepiece or introduction of foreign objects. Courtesy of General Electric Research Laboratory.

found a wide variation between different facepieces, respiratory efficiency, and protection factors. Protection factors were noted as low as 5 (for single use and quarter masks for dusts) to as high as 10,000 for an open-circuit/pressure-demand/full-facepiece SCBA. The protection factors from the ANSI Z88.2-1980 standard are shown in Table 6.7.

Fit Tests

Two types of fit tests have been recognized. In the qualitative test, a person wearing a respirator is exposed to an irritant smoke (such as titanium tetrachloride smoke), an odorous vapor (such as isoamyl acetate vapor), or other test agent. An air-purifying respirator must be equipped with one or more air-purifying elements which effectively remove the test agent from inside the facepiece. If the wearer is unable to detect penetration of the agent into the facepiece, the wearer is judged as having a satisfactory fit. It should be noted that persons vary widely in their ability to

TABLE 6.7. Respirator Protection Factors[a]

Type of Respirator	Permitted for Use in Oxygen-Deficient Atmospheres	Permitted for Use in Atmospheres Which are Immediately Dangerous to Life or Health[b]	Respirator Protection Factor	
			Qualitative Test	Quantitative Test
Particulate-filter, quarter-mask or half-mask facepiece[b,c,d]	No	No	10	As measured on each person with maximum of 100 if dust, fume, or mist filter is used, or maximum of 3000 if high-efficiency filter is used, whichever is less
Vapor or gas removing, quarter-mask or half-mask facepiece[d]	No	No	10, or maximum use limit of cartridge or canister for vapor or gas, whichever is less	As measured on each person with maximum of 3000, or maximum use limit of cartridge or canister for vapor or gas, whichever is less
Combination particulate-filter and vapor- or gas-removing, quarter-mask or half-mask facepiece[c,d]	No	No	10, or maximum use limit of cartridge or canister for vapor or gas, whichever is less	As measured on each person with maximum of 100 if dust, fume, or mist filter is used, maximum of 3000 if high-efficiency filter is used, or maximum use limit of cartridge or canister for vapor or gas, whichever is less
Particulate-filter, full facepiece[c]	No	No	100	As measured on each person with maximum of 100 if dust, fume, or mist filter is used, or maximum of 3000 if high-efficiency filter is used, whichever is less

TABLE 6.7. *(Continued)*

Type of Respirator	Permitted for Use in Oxygen-Deficient Atmospheres	Permitted for Use in Atmospheres Which are Immediately Dangerous to Life or Health[b]	Respirator Protection Factor	
			Qualitative Test	Quantitative Test
Vapor- or gas-removing, full facepiece	No	Yes, if concentration of hazardous substance does not exceed product of respirator PF and permissible concentration, and/or maximum use limit of cartridge or canister for vapor or gas	100, or maximum use limit of cartridge or canister for vapor or gas, whichever is less	As measured on each person with a maximum of 3000, or maximum use limit of cartridge or canister for vapor or gas, whichever is less
Combination particulate-filter and vapor- or gas-removing, full facepiece[c]	No	No	100, or maximum use limit of cartridge or canister for vapor or gas, whichever is less	As measured on each person with maximum of 100 if dust, fume, or mist filter is used, maximum of 3000 if high-efficiency filter is used, or maximum use limit of cartridge or canister for vapor or gas, whichever is less
Powered particulate-filter, any respiratory-inlet covering[c,d,e]	No	No (yes, if escape provisions are provided[e])	n.a.[f]	n.a.
			(No tests are required due to positive-pressure operation of respirator. The maximum protection factor is 100 if dust, fume, or mist filter is used, or 3000 if high-efficiency filter is used, whichever is less.)	

Type					
Powered vapor or gas removing, any respiratory-inlet covering[c,d,e]	No	No (yes, if escape provisions are provided[e])	n.a.	(No tests are required due to positive-pressure operation of respirator. The maximum protection factor is 3000, or maximum use limit of cartridge or canister for vapor or gas, whichever is less.)	n.a.
Powered combination particulate-filter and vapor-or gas-removing, any respiratory-inlet covering[c,d,e]	No	No (yes, if escape provisions are provided[e])	n.a.	(No tests are required due to positive-pressure operation of respirator. The maximum protection factor is 100 if dust, fume, or mist filter is used, 3000 if high-efficiency filter is used, or maximum use limit of cartridge or canister for vapor or gas, whichever is less.)	n.a.
Air-line, demand, quarter-mask or half-mask facepiece, with or without escape provisions[d,g]	No	No	10	As measured on each person with maximum of 3000, whichever is less	
Air-line, demand, full facepiece, with or without escape provisions[d,g]	No	No	100	As measured on each person with maximum of 3000, whichever is less	
Air-line, continuous-flow or pressure-demand, any facepiece, without escape provisions[d]	No	No	n.a.	(No tests are required due to positive-pressure operation of respirator. The maximum protection factor is 3000.)	

TABLE 6.7. (Continued)

Type of Respirator	Permitted for Use in Oxygen-Deficient Atmospheres[b]	Permitted for Use in Atmospheres Which are Immediately Dangerous to Life or Health[b]	Respirator Protection Factor	
			Qualitative Test	Quantitative Test
Air-line, continuous-flow or pressure-demand, any facepiece, with escape provisions[a,g]	Yes[b]	Yes	n.a.	n.a.
			(No tests are required due to positive-pressure operation of respirator. The maximum protection factor is 10,000 plus.[h])	
Air-line, continuous-flow, helmet, hood, or suit, without escape provisions	No	No	n.a.	n.a.
			(No tests are required due to positive pressure operation of respirator. The maximum protection factor is 3000.)	
Air-line, continuous-flow, helmet, hood, or suit, with escape provisions[g]	Yes[b]	Yes	n.a.	n.a.
			(No tests are required due to positive-pressure operation of respirator. The maximum protection factor is 10,000 plus.[h])	
Hose mask, with or without blower, quarter-mask or half-mask facepiece[d]	No	No	10	As measured on each person with maximum of 3000, whichever is less
Hose mask, with or without blower, full facepiece	No	No	100	As measured on each person with maximum of 3000, whichever is less
SCBA, open-circuit demand or closed-circuit, quarter-mask or half-mask facepiece[d]	Yes	No	10	As measured on each person with maximum of 3000, whichever is less

SCBA, open-circuit demand or closed-circuit, full facepiece or mouthpiece/nose clamp[d]	Yes[b]	No (yes, if respirator is used for mine rescue)	100	As measured on each person with maximum of 3000, whichever is less
SCBA, open-circuit pressure-demand, quarter-mask or half-mask facepiece, full facepiece, or mouthpiece/nose clamp[d]	Yes[b]	Yes	n.a.	n.a.
(Combination respirators not listed.)	(No tests are required due to positive-pressure operation of respirator. The maximum protection factor is 10,000 plus.[h])	(The type and mode of operation having the lowest respirator protection factor shall be applied to the combination respirator.)		

Source. "Standards for Respiratory Protection," ANSI Z88.2-1980.

[a] A respirator protection factor is a measure of the degree of protection provided by a respirator to a respirator wearer. Multiplying the permissible time-weighted-average concentration or the permissible ceiling concentration, whichever is applicable, for a toxic substance, or the maximum permissible airborne concentration for a radionuclide by a protection factor assigned to a respirator gives the maximum concentration of the hazardous substance for which the respirator can be used. Limitations of filters, cartridges, and canisters used in air-purifying respirators shall be considered in determining protection factors.

[b] See definition for oxygen deficiency—immediately dangerous to life or health.

[c] When the respirator is used for protection against airborne particulate matter having a permissible time-weighted-average concentration less than 0.05-mg particulate matter per m³ of air or less than 2 million particles per ft³ of air, or for protection against airborne radionuclide particulate matter, the respirator shall be equipped with a high-efficiency filter(s).

[d] If the air contaminant causes eye irritation, the wearer of a respirator equipped with a quarter-mask or mouthpiece/nose clamp shall be permitted to use a protective goggle or a respirator equipped with a full facepiece.

[e] If the powered air-purifying respirator is equipped with a facepiece, the escape provisions means that the wearer is able to breath through the filter, cartridge, or canister and through the pump. If the powered air-purifying respirator is equipped with a helmet, hood, or suit, the escape provisions shall be an auxiliary self-contained supply of respirable air.

[f] n.a. = not applicable.

[g] The escape provisions shall be an auxiliary self-contained supply of respirable air.

[h] The protection-factor measurement exceeds the limit of sensitivity of the test apparatus. Therefore the respirator has been classified for use in atmospheres having unknown concentrations of contaminants.

Note. Respirator protection factors for air-purifying type respirators equipped with a mouthpiece/nose clamp form of respiratory-inlet covering are not given since such respirators are approved only for escape purposes.

sense or smell odors, and this factor should be considered before giving a complete endorsement of the qualitative test.

In the quantitative fit test, which is the test recommended by the NIOSH, the wearer enters an atmosphere containing a test aerosol or gas, such as polydisperse sodium chloride aerosol, polydisperse DOP (dioctylphthlate), or dichlorodifluoro-methane (F-12). Questions have been raised about the safety of the use of DOP aerosol, and it is likely this test agent will be replaced. Corn-oil aerosol should be substituted for di-2-ethylhexylphthalate (DEHP) as an agent for measuring facepiece leakage in respiratory-fit testing, in view of data indicating that DEHP is a potential carcinogen, according to a recent hazard review by the NIOSH. Of three possible DEHP substitutes examined, corn oil was the only one for which low toxicity was well documented.

Test chambers and detailed procedures for both tests are detailed in ANSI Z88.2-1980, Appendix A6. The exercises used ensure that the respiratory device is, in fact, providing proper protection under all conditions likely to be encountered at work.

Education and Training

It is recognized that respirators are not a comfortable addition to the face.[29] The importance of the respirator as temporary protection is significant. If the wearer is in good physical condition, and no respiratory or cardiovascular deficiencies exist, and the facepiece is properly fitted by tests, day-by-day implementation requires full understanding of the advantages of using the device. If the facial hair, such as the male beard, is allowed to grow more than 2 days (the average facial hair grows about $1/72$ in./day), a proper seal may be difficult, if not impossible. Other excessive facial or head hair, including hairdos which interfere with the proper use of respiratory protective devices, must be avoided.

6.4.   OTHER FACTORS AND CONSIDERATIONS

In implementing and maintaining an effective respiratory protection program, in addition to selecting, fitting, and maintaining the proper equipment for a particular work environment, other factors may be important. For example:

The cost differences between various types of devices.

The length of time a device must be worn and its relationship to the wearer's comfort and acceptance.

Availability of breathing air of known certified purity for air-line use, and for the refilling of cylinders in the SCBA.

The coordination of the respiratory protection with other personal protective equipment needed for the job, such as face shield, welding goggles, earmuffs, and hard hats.

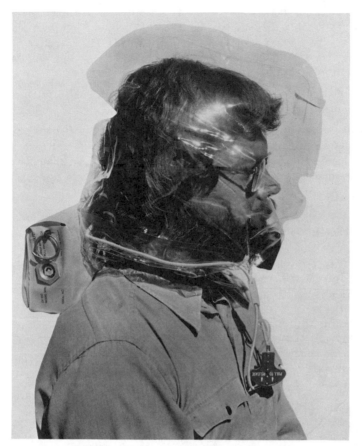

Figure 6.13. Escape mask/hood. Such devices have an effective life of five minutes or less and may be used for escape from confined spaces, as in the engine rooms of ships. Courtesy of Robertshaw.

Temperature extremes which may make it desirable to supply warmed or cooled air to the respiratory-inlet covering.

Use of prescription lenses by workers who need proper corrective lenses to perform their assigned tasks.

Stability of the respiratory-inlet covering on the face while performing certain movements; for example, a half-mask respirator with two-point strap suspension is more unstable on the face than a mask with four-point suspension and a full facepiece is more stable than a quarter- or half-mask.

Cartridge or canister capability and length of time it may be used before contaminant breakthrough can occur. (If test data are not available, the manufacturer of the device should be required to furnish it or an air-supplied device used instead).

Communication between workers. The use of in-mask microphones or throat masks external to the mask should be considered, and either amplifiers or two-way radio equipment used where necessary [see H. H. Fawcett, "Speech Transmission Through Respiratory Protective Devices," *Am. Ind. Hyg. Assoc. J.* **22,** No. 3, 170–174 (June 1961).

Physical limitations and capabilities of each individual user, including work requirements versus physical characteristics.

Vision restriction, mobility, and other safety factors. Provision for escape, as from the engine room of a ship, in case of serious fire emergency, should be considered. The mask/hood shown (see Fig. 6.13) is an example of an escape device intended to be donned in seconds, and to give a 5-min air supply for escape purposes.

REFERENCES

1a.   "The Radon Question: Government, Independent Studies Show Home Insulation May be Hazardous," *Consumers' Research Magazine* **63,** No. 11 12–14 (November 1980).

   b.   Tracketch Radon in water measurement services, June 1983, Terradex Corporation, Walnut Creek, CA 94598.

   c.   "The Air You Breathe," *EPA J.* **4,** No. 9 (October 1978), (U.S. Environmental Protection Agency, Washington, D.C. 20460.)

   d.   "Bromine in Arctic Air: Spring Peak May Destroy Ozone Layer," *Chem. Eng. News* 6 (August 22, 1983).

2.   J. Dorigan, B. Fuller, and R. Duffy, "Scoring of Organic Air Pollutants: Chemistry, Production and Toxicity of Selected Synthetic Organic Chemicals," MITRE Technical Report MTR-7248, September 1976, U.S. Environmental Protection Agency, Washington, D.C.

3.   B. G. Ferris, Jr., "Literature Review of Health Effects of Air Pollution," American Thoracic Society, American Lung Association, New York, 1973.

4.   J. E. McFadden, J. H. Beard, III, and D. Moschendreas, "Survey of Indoor Air Quality, Health Criteria and Standards," Final Report, GEOMET EF-595, August 1977, U.S. Environmental Protection Agency and U.S. Housing and Urban Development, Washington, D.C.

5.   J. H. Comroe, *Physiology and Respiration,* Yearbook Medical Publishers, Chicago, 1965.

6.   "An Investigation of Self-Contained Breathing Apparatus for Use in Mines," Report of a Committee of the South Midland Coal Owners upon an Investigation Conducted in the Mining Department in the University of Birmingham into Self-Contained Breathing Apparatus for Rescue and Recovery Work in Mines after Underground Fires and Explosions, Cornish Brothers Ltd., Birmingham, England, 1910–1911.

7.   Marshall Sittig (Ed.), *Priority Toxic Pollutants—Health Effects and Allowable Limits,* Noyes Data Corp., Park Ridge, N.J., 1980.

8.   Marshall Sittig, (Ed.), *Hazardous and Toxic Effects of Industrial Chemicals,* Noyes Data Corp., Park Ridge, N.J., 1979.

9.   "Principles of Toxicological Interactions Associated with Multiple Chemical Exposures," Panel on Evaluation of Hazards Associated with Maritime Personnel Exposed to Multiple Cargo Vapors, Board on Toxicology and Environmental Health Hazards and Committee on Maritime Hazardous Materials, National Research Council, Washington, D.C., 1980.

10. "Particulate Polycyclic Organic Matter, Division of Medical Sciences." NTIS No. PB 212 940, 1972, Committee on Biologic Effects of Atmospheric Pollutants, National Research Council and U.S. Environmental Protection Agency, Washington, D.C.

11. John M. Dement, "Estimates of Pulmonary and Gastrointestinal Deposition for Occupational Fiber Exposures," DHEW (NIOSH) Publication No. 79-135, April 1979.

12. L. R. Ember, "Acid Rain Focus of International Cooperation," *Chem. and Eng. News,* 15–17 (December 3, 1979).

13. J. E. McFadden and M. D. Koontz, "Sulfur Dioxide and Sulfates Materials Damage Study," GEOMET ES-812, February 1980, U.S. Environmental Protection Agency, Research Triangle Park, N.C.

14. Ruth Porter (Ed.), "Breathing: Hering-Breuer Centenary Symposium," A CIBA Foundation Symposium, Churchill, London, 1970.

15. G. R. C. Atherley, *Occupational Health and Safety Concepts,"* Applied Science Publishers, London, 1978, pp. 25–55.

16. H. H. Fawcett, "Hydrogen Sulfide—Killer That May Not Stink," *J. Chem. Ed.* **9,** No. 25, 511 (1948).

17. R. P. Smith, and R. E. Gosselin, "Hydrogen Sulfide Poisoning," *J. Occup. Med.* **21,** No. 2, 93–97 (1979).

18a. National Academy of Sciences/National Research Council, "Carbon Monoxide: Effects of Chronic Exposure to Low Levels of CO on Human Health, Behavior and Performance," Washington, D.C., 1969.

  b. S. A. Lindgren, "A Study of the Effect of Protracted Occupational Exposure to Carbon Monoxide," *Acta Med. Scand. Stockholm,* Supplement 356 (1961).

  c. Y. Sawada, "Correlation of Pathological Findings with Computed Tomographic Findings, After Acute Carbon Monoxide Poisoning," *New Eng. J. Med.* **308,** No. 21, 1296 (May 26, 1983).

19. W. H. Stahl (Ed.), "Compilation of Odor and Taste Threshold Values Data," DS 48, Publication Code Number 05-048000-36, 1977, American Society for Testing and Materials, Philadelphia, Pennsylvania.

20. T. Temple, "On the Cutting Edge," *EPA J.* **6,** No. 9, 12–15 (October 1980).

21. A. D. Little, "Recommended Industrial Ventilation Guidelines," (NIOSH) 76-162, January 1976, National Institute for Occupational Safety and Health, Cincinnati, Ohio.

22. W. E. Ruch and B. J. Held, *Respiratory Protection: OSHA and the Small Businessman,* Ann Arbor Science Publishers, Ann Arbor, 1975.

23. D. D. Douglas, "Respiratory Protective Devices," in G. D. Clayton and F. E. Clayton (Eds.), *Patty's Industrial Hygiene and Toxicology,* Vol. 1, 3rd rev. ed., Wiley–Interscience, 1978, pp. 993–1057.

24. "Air, Compressed for Breathing Purposes," Federal Specification BB-A-1034a, June 21, 1968.

25. "Breathing Apparatus, Self-Contained," Interim Federal Specification GG-B-675d, September 23, 1976.

26. "Oxides, Oxygen Producing," Military Specification MIL-0-1563c, September 25, 1964.

27. "Emergency Oxygen Supply, Chlorate Candle," Military Specification MIL-E-83252, Aircraft CRU-74/P, February 20, 1970.

28. E. C. Hyatt, "Respirator Protection Factors," Los Alamos Scientific Laboratory Report No. LA-6084-MS, January 1976, Los Alamos, California.

29. "Bibliography of Journal Articles, Reports and Other Publications on Physiology and Psychology of Respirator Use," International Respirator Research Workshop, September 8 1980–September 11, 1980, National Institute for Occupational Health and Safety, Morgantown, West Virginia.

BIBLIOGRAPHY

"Air Pollutants NOx and Ozone" (Editorial), *Env. Sci. Tech.* **15,** No. 3 253–254 (March 1981).

"American National Standard Practices for Respiratory Protection," ANSI Z88.2-1980, and "Respiratory Protection for the Fire Services," ANSI Z88.5-1980, American National Standards Institute, New York, 1980.

Anon., "Breathing New Life into Respiratory Protection," *Occup. Health Safety,* 44–47 (August 1982).

Anon., "Chemical-Biological Warfare in Afghanistan," *Wall Street Journal,* June 7, 1982, p. 20.

Anon., "OSHA to Seek Revisions in Cotton Dust Standards," *Washington Post,* June 8, 1983, p. A-23; also *see Wall Street Journal,* June 8, 1983, p. 10. (Proposed regulations would exempt knitting industry and other industries from 1978 standard which would save $94 million in capital costs and $30.7 million annually.)

Asbestos, by Schneider, A. L., *Proc. of the Marine Safety Council,* Vol. 41, No. 1, pp. 75–76, March 1984.

Asbestos: Properties, Application and Hazards, Ed. by S. S. Chissick and R. Derricott, Vol. 1, 1979; Vol. 2, 1983, Wiley-Interscience, Chichester, U.K. See also "U. S. Court Rejects Emergency Limit on Asbestos Levels," *Wall Street Journal,* March 9, 1984, p. 14. (Court rejected reduction of exposure level from 2 million to 500,000 fibers per cubic meter of air.)

Bidstrup, P. L., *Toxicity of Mercury and Its Compounds,* Elsevier, Amsterdam, 1964.

Birkner, L. R., *Respiratory Protection: A Manual and Guideline,* Celanese Corporation, American Industrial Hygiene Association, Akron, Ohio, 1980.

Braker, W., and Mossman, A. L., "Effects of Exposure to Toxic Gases—First Aid and Medical Treatment," Matheson Gas Products, East Rutherford, N.J., 1970.

Buchanan, W. D., *Toxicity of Arsenic Compounds,* Elsevier, Amsterdam, 1962.

Burrows, B. *Respiratory Insufficiency,* Yearbook Medical Publishers, Chicago, 1975.

"Clean Air Realism" (Editorial), *Wall Street Journal,* December 31, 1981, p. 6.

Comroe, J. H., Jr., *Lung: Clinical Physiology and Pulmonary Function Tests,* 2nd ed., Yearbook Medical Publishers, Chicago, 1962.

Constance, J. D., *Controlling In-Plant Airborne Contaminants: System Design and Calculations*, Marcel Dekker, Inc., N.Y., 1984.

Cotes, J. E., *Lung Function,* 3rd ed., Lippincott, Philadelphia, 1975.

Crofton, J., and Douglas, A., *Respiratory Diseases,* 2nd ed., Lippincott, Philadelphia, 1975.

Ebling, P. R., "Phosphorus, A Respiratory Poison," Dissertation, University of Cincinnati, Cincinnati, Ohio, 1960.

Ferris, B. G., Jr., "Literature Review of Health Effects of Air Pollution," American Thoracic Society, American Lung Association, New York, 1973.

Florent (nee Jarlet), Patricia, "Intoxications par les vapeurs de fuel lourd, Thèse pour le doctcrat en Medecine," University of Paris, 1976.

Flury, F., and Zernick, G., *Schadliche Gase* (Noxious Gases, Vapors, Mist, and Smoke-and-Dust Particles), Springer-Verlag, Berlin, 1931.

Gerarde, H. W., *Toxicology and Biochemistry of Aromatic Hydrocarbons,* Elsevier, Amsterdam, 1960.

Geschickter, C. F., *Lung in Health and Disease,* Lippincott, Philadelphia, 1973.

Glass, N. R., Effects of Acid Precipitation, *Eng. Sci. Tech.* **16,** No. 3, 162A–169A (March 1982). See also *An Updated Perspective on Acid Rain,* Edison Electric Institute, Washington, D.C., 1981.

Goldwater, L. J., *Mercury, A History of Quicksilver,* York Press, Baltimore, 1972.

Hack, A., Hyatt, E. C., Held, B. J., Moore, T. O., Richards, C. P., and McConville, J. T., "Selection of Respirator Test Panels Representative of U.S. Adult Facial Sizes," Los Alamos Scientific

Laboratory Report No. LA-5488, U.C. 41, reporting date December 1973, issued March 1974, Los Alamos, N.M.

"Handbook of Hazardous Materials," Technical Guide No. 7, American Mutual Insurance Alliance, Chicago, Illinois, 1983.

Heitzman, E. R., *Lung Radiologic–Pathologic Correlations,* Mosby, St. Louis, 1973.

Henderson, Y., and Haggard, H. W., *Noxious Gases,* American Chemical Society Monograph Series, American Chemical Society, Washington, D.C., 1943.

Hepple, P., *Lead in the Environment,* Institute of Petroleum, London, England, 1972.

Henry, Norman W., III, "Respiratory Cartridge and Canister Efficiency Studies with Formaldehyde," *Am. Ind. Hyg. Assoc. J.* **42,** No. 12, 853–857 (December 1981).

Hunsinger, J., *Respiratory Technology,* 2nd ed., Reston, Va., 1976.

Hyatt, E. C., and White, J. M., "Respirators and Protective Clothing," Safety Series No. 22, International Atomic Energy Agency, 1967.

Hygienic Guides, American Industrial Hygiene Association, Akron, Ohio, 1964–1980.

"Industrial Safety Data Sheets on Specific Chemicals," National Safety Council, Chicago, Ill., 1982.

"Instrumentation for Environmental Monitoring, AIR, Part 2," 1st ed., Environmental Instrumentation Group, LBL-1, Vol. 1, Part 2, Lawrence Berkeley Laboratory, University of California, Berkeley, 1975.

International Association of Fire Fighters, *Life Support, A Fire Guide To Self-Contained Breathing Apparatus,* Washington, D.C., 1981.

International Society for Respiratory Protection, P.O. Box 7567, St. Paul, MN 55119.

"International Symposium, Environmental Health Aspects of Lead," Proceedings, Amsterdam, October 2, 1972–October 6, 1972, Commission of the European Communities, CID, Luxembourg, May 1973.

"Iraq's Chemical War" (Editorial), *Washington Post,* March 11, 1984, p. C6, and "New 'Yellow Rain' Victims," *Wall Street Journal,* March 12, 1984, p. 28, and "The Trail to Crazy Island" (Review and Outlook), *Wall Street Journal,* March 16, 1984, p. 26.

James, R. H., "Breathing Resistance and Dead Space in Respiratory Protective Devices," DHEW (NIOSH) Publication No. 77-161, October 1976.

Junod, A. F., and DeHaller, R., "Lung Metabolism: Proteolysis and Antiproteolysis, Biochemical Pharmacology, Handling of Bioactive Substances," *Proceedings of International Symposium,* Davos, Switzerland, October 1974, Academic Press, New York, 1976.

Kamon, E., "Cooling Efficiency of Different Air Velocities in Hot Environments," DHEW (NIOSH) Publication No. 79-129, March 1979.

Labows, J. N., Jr., "What the Nose Knows: Investigating the Significance of Human Odors." *The Sciences,* 10–13 (November 1980).

"Lead Environmental Health Criteria 3, "World Health Organization, Geneva, Switzerland, 1977.

Lead versus Health: Sources, and Effects of Low Level Lead Exposure, Ed. by M. Rutter and R. R. Jones, Wiley, Chichester, U.K. 1983.

Lee, K. P., and Trochimowicz, H. J. "Induction of Nasal Tumors in Rats Exposed to Hexamethyl-phosphoramide (HMPA) by Inhalation," *J. Nat. Cancer Inst.* **68,** No. 1, 157–171 (January 1982).

Lederer, W. H., and Fensterheim, R. J., eds., *Arsenic-Industrial, Biomedical, Environmental Perspectives,* Van Nostrand Reinhold, N.Y., 1984.

Leithead, C. S., and Lind, A. R., *Heat Stress and Heat Disorders,* F. A. Davis, Philadelphia, 1974.

Lindsay, D. B., and Stricoff, R. S., *A Feasibility Study of the State-of-the Art of Personnel Monitors,* Arthur D. Little, Inc., Cambridge, Mass., 1978. (DOT-CG-73211-A, NTIS Number AD A 072992, National Technical Information Service, Springfield, Virginia, 1978.

Matheson Guide to Safe Handling of Compressed Gases, Matheson Division Searle Medical Products, Secaucus, N.J., 1982.

"Mercury, Environmental Health Criteria 1," World Health Organization, Geneva, Switzerland, 1976.

Moeschlin, S., "Outstanding Symptoms of Poisoning," in *Diagnosis and Treatment,* Grune & Stratton, New York, 1965, pp. 644–678.

NAS-NRC, "Physiological and Toxicological Aspects of Combustion Products," International Symposium, March 18, 1974—March 20, 1974, Committee on Fire Research, National Research Council, Washington, D.C., 1976.

"Nitrogen, Data Sheet," Chemical Section, National Safety Council, Chicago, Ill., 1982.

"Oxides of Nitrogen," Environmental Health Criteria 4, World Health Organization, Geneva, Switzerland, 1977.

Padour, J., and Shaw, A., "Respiratory Testing," pp. 16–19; Horvath, E., "Respiratory Testing," pp. 20–31; Held, B. J., and Richards, C. B., "Respiratory Protection," pp. 32–35; Murphy, A. J., "Respiratory Testing Role of OHN," pp. 36–40, *Occup. Health Safety* **46,** No. 5 (September/October 1977).

Pesticides Inspection Manual, Pesticides Enforcement Division U.S. Environmental Protection Agency, Washington, D.C., 1980.

"Practices for Respiratory Protection During Fumigation," ANSI Z88.3, Draft August 18, 1981, American National Standards Institute New York, 1981.

Proctor, N. H., and Hughes, J. P., *Chemical Hazards of the Workplace,* Lippincott, Philadelphia, 1978.

"Respirator Studies for the National Institute for Occupational Safety and Health, January 1–December 31, 1977," Los Alamos Scientific Laboratory Report No., UC-41, LA-7317-PR, DHEW (NIOSH) Publication No. 78-161, June 1978.

"Respiratory Protection," Occupational Safety and Health Program, U.S. Environmental Protection Agency, Washington, D.C. 1980.

Rekus, J. F., *Evaluation of Worker Exposure to Airborne Chemical Agents,* American Chemical Society National Meeting, Washington, D.C. Aug. 29, 1983.

"Respiratory Protection Against Radon Daughters," ANSI Z88.1-1975, American National Standards Institute, New York, 1975.

"Respiratory Protective Devices," Australian Standards CZ 11 and Z 18-1968, Standards Association of Australia, North Sydney, New South Wales, 1968.

"Respiratory Protective Equipment" and "Respiratory System," in *Encyclopedia of Occupational Health & Safety,* McGraw-Hill, New York, 1972, pp. 1214–1220 and 1220–1223.

"Respiratory Protective Equipment," Information Sheet No. 9, International Labour Office, Geneva, Switzerland 1964.

"Respiratory Cancer, Occupational," p. 1916–1919; "Respiratory Protective Equipment," pp. 1919–1926; also "Respiratory System," pp. 1926–1930, in *Encyclopaedia of Occupational Health & Safety,* 3rd Rev. Ed., International Labour Organization, Geneva, 1983.

"Safety Precautions for Oxygen, Nitrogen, Argon, Helium, Carbon Dioxide, Hydrogen, Acetylene, Ethylene Oxide, and Sterilant Mixtures," Publication F-3499C, March 1976, Linde Division, Union Carbide Corp., New York.

Staub, N. C., *Lung Water and Solute Exchange* (Lung Biology in Health and Disease Series), Dekker, New York, 1978.

Stern, A. C., *Air Pollution,* 2nd ed. (in 3 volumes), Academic Press, New York, 1968.

Strauss, M. J., *Lung Cancer: Clinical Diagnosis and Treatment, Clinical Oncology Monograph,* Grune & Stratton, New York, 1977.

Sykes, M. K., *Respiratory Failure,* 2nd ed., Lippincott, Philadelphia, 1976.

"Time, Air, and Money" (Review and Outlook), *Wall Street Journal,* 6 March 3, 1982, p. 26.

Tuck, A., Johnson, J. W., and Moulton, D. C., *Human Responses to Environmental Odors*, Academic Press, New York, 1974.

WHO Scientific Group, "Respiratory Viruses," Technical Report Series No. 408, World Health Organization, Geneva, 1969.

"Yellow Rain: Hmong, Afghans, Now Iranians," *Wall Street Journal*, March 13, 1984, p. 30.

"Yellow Rain," by L. Ember, *Chem. & Engr. News*, pp. 8–34, Jan. 9, 1984; See also "Science and Windmills" (Review and Outlook), *Wall Street Journal*, Feb. 15, 1984, p. 30, and Letters to Editor, *Wall Street Journal*, p. 35, March 6, 1984, and Editorial, p. 26, March 16, 1984.

Publications of the National Academy of Sciences—National Research Council, *Washington, D.C.*

"Carbon Monoxide: Effects of Chronic Exposure to Low Levels of CO on Human Health, Behavior, and Performance" 1969.

"Chlorine and Hydrogen Chloride," 1976.

"Chromium," 1974.

"Fluorides," 1971.

"Lead—Airborne Lead in Perspective," 1972.

"Manganese," 1973.

"Nickel," 1975.

NO_x(Nitrogen Oxides) Emission Controls for Heavy-Duty Vehicles: Toward Meeting a 1986 Standard, 1981.

"Selenium," 1976.

Publications of the National Institute for Occupational Safety and Health, *Cincinnati, Ohio*

"Abrasive Blasting Respiratory Protective Practices," NIOSH 74–104, 1974.

"An Air-Supplied Respirator for Underground Coal Miners," CONTR HSM-99-71-43, 1976.

"An Evaluation of Organic Vapor Respirator Cartridges and Canisters Against Vinyl Chloride," NIOSH 75–111, 1975.

"Anthropometry for Respirator Sizing," CONTR HSM-99-71-11, 1977.

"Breathing Resistance and Dead Space in Respiratory Protective Devices," NIOSH 77–161, 1977.

"Design Specifications for Respiratory Breathing Devices for Firefighters," NIOSH 76–121, 1976.

"Determination of Respirator Filtering Requirements in a Coke Oven Atmosphere," CONTR CDC-99-74-88, 1980.

"Development of a Prototype Service Life for Organic Vapor Respirators," NIOSH 78–170, 1978.

"Development of Improved Respirator Cartridge and Canister Test Methods," NIOSH 77–209, 1977.

"Engineering Control of Welding Fumes," NIOSH 75–115, 1975.

"Evaluation of the NIOSH Certification Program Division of Safety Research Testing and Certification Branch," NIOSH 80–113, 1980.

"Evaluation of Two-Way Valves for Respiratory Testing," NIOSH 75–123, 1975.

"Evaluation of Two-Way Valves for Resting Level Respiratory Testing," NIOSH 77–212, 1977.

"Exhalation Valve Leakage Test" NIOSH 78–112, 1978.

"Human Variability and Respirator Sizing," NIOSH 76–146, 1976.

"Industrial Health and Safety Criteria for Abrasive Blast Cleaning Operations," NIOSH 75–122, 1975.

"NIOSH Certified Equipment List as of Sept. 1," NIOSH 83–122 (revised annually), 1983.

"Performance Evaluation of Respiratory Protective Equipment Used in Paint Spraying," NIOSH 78–177, 1978.

Respirator User's Notices: (1) *Effects of Chemicals on Rubber and Plastic Parts of Self-Contained Breathing Apparatus*; (2) *Effects of Heat and Flames on Rubber and Plastic Parts of Self-Contained Breathing Apparatus*; and (3) *Provision of New Elbow Fitting for MSA Self-Contained Breathing Apparatus*, by T. C. Purcell, December 16, 1983. Available from NIOSH Safety and Health—ALOSH, Morgantown, WV 26505.

"Respiratory Protection—A Guide for the Employee," NIOSH 78–193B, 1978.

"Respiratory Protection—An Employer's Manual," NIOSH 78–193A, 1978.

"Suggested Research and Development Programs for Gas and Vapor Respirators," CONTR 210-73-0080, 1978.

"Survey of Personal Protective Equipment Used in Foundries," NIOSH 80–100, 1980.

7

RCRA, SUPERFUND, and Guides to Their Implementation

Waste is a byproduct of most research and production, but until quite recently has received relatively little attention. Some wastes are hazardous, but many are not, and the disposal has not always been conducted in a manner which considered the effects, or potential effects, on humans, the environment, the biota, and the long-term economy of the area.

Resource Conservation and Recovery Act (RCRA)

Several incidents came to public attention during the 1970s which created the demand for a more organized approach to waste. A glimpse of poor practices for the handling of wastes is shown in Fig. 7.1. Congress responded to this demand for adequate control and utilization of discarded materials and wastes by passing the Resource Conservation and Recovery Act (RCRA) (P.L. 94-580) on October 21, 1976.[1] This law, which is cited as 42 USC 6901, the Solid Waste Disposal Act of 1976, gave the incentive to those generating, transporting, treating, storing, and disposing of solid waste to consider carefully their management practices. (As specified in the law, the wastes may be gaseous, liquid, or solid, or any variation of the physical state.) The RCRA directed the EPA to promulgate regulations to protect human health and the environment from improper management of hazardous (Subtitle C) and nonhazardous wastes (Subtitle D). It further provides technical and financial assistance for the development of management plans and facilities for the recovery of energy and other resources.

Under Subtitle C of the RCRA, the EPA is given the authority to impose a

Figure 7.1. "Real-world" practices of hazardous waste disposal which are no longer acceptable, or legal. Courtesy of U.S. EPA.

"cradle to grave" control system for hazardous waste. The EPA has defined hazardous waste using the dual approaches of identifying general characteristics and listing specific hazardous waste, sources, and process waste streams. The law establishes the minimum amount of waste generated on a monthly basis that will require adherence to the RCRA regulations. Any facility that stores (for more than 90 days), processes, or disposes of hazardous waste must meet the design and operating standards set forth in the regulations. Minimum performance standards are also proposed for incineration processes. An important feature of the act is that it requires the complete tracking of hazardous waste from its point of generation, through each step of processing to the actual disposal of the waste. The tracking will be accomplished by a manifest system that will document each movement of the waste until it is ultimately disposed of. Table 7.1 notes several pertinent sections of the RCRA regulations and the date on which they were promulgated by publication in the *Federal Register*.

Subtitle D of the law requires an inventory of all nonhazardous-waste-disposal sites. Under the regulations promulgated for Subtitle D, all open dumps will be

TABLE 7.1. Relevant Portions of the RCRA and Their Status

RCRA Section	Federal Register Date	Status
Section 3001		
Identification and listing of hazardous waste	5/19/80	Promulgated
Section 3002		
Standards applicable to generators of hazardous waste	2/26/80	Promulgated
Section 3003		
Standards applicable to transporters of hazardous waste	2/26/80	Promulgated
Section 3004		
Standards applicable to owners and operators of hazardous waste treatment, storage, and disposal facilities	5/19/80	Promulgated
Section 3004		
Interior status standards applicable to owners and operators of hazardous waste treatment, storage, and disposal facilities	5/19/80	Promulgated
Section 3005		
Permits for treatment, storage and disposal of hazardous waste	5/19/80	Promulgated
Section 3006		
Guidelines for authorized state hazardous waste programs	5/19/80	Promulgated
Section 3010		
Preliminary notification of hazardous waste activity	2/26/80	Promulgated

Additional Final Rules and Regulation Interpretation
(With Date of Publication in Federal Register)

40 CFR Part 122, November 10, 1980
Consolidated Permit Regulations and Hazardous Waste Management System
40 CFR 261, November 12, 1980
Hazardous Waste Management System: Identification and Listing of Hazardous Waste—Finalizing the Lists of Hazardous Wastes (261.31 and 261.32) and Proposal to Amend. 261.32
40 CFR 122, 260, 264, and 265, November 17, 1980
Hazardous Waste Management System: Suspension of Rules and Proposal of Special Standards for Wastewater Treatment Tanks and Neutralization Tanks
40 CFR 261, November 19, 1980
Hazardous Waste Management System: Mining and Cement Kiln Wastes Exemptions; Small Quantity Generator Standards; Generator Waste Accumulation Amendment; Hazardous Waste Spill Response Exception, and clarification of interim status requirements
40 CFR 261, 262, and 265, November 25, 1980
Hazardous Waste Management System: Clarification of Regulations on Hazardous Waste in Containers; Exception of Certain Treated-Wood Wastes; Final List of Commercial Products Which are Hazardous Wastes if Discarded (S261.33); Exclusions in Response to Delisting Petitions
Hazardous Waste Management System: Addition of General Requirements for Treatment, Storage and Disposal Facilities (40 CFR Part 264); Amendment of Interim Status Standards Respecting Closure and Post-Closure Care and Financial Responsibility (40 CFR Part 265) and Conforming Amendments to the Permitting Requirements (40 CFR Part 122), *Fed. Reg.* **46,** No. 7, 2802–2892 (January 12, 1981)

TABLE 7.1. (*Continued*)

Additional Final Rules and Regulation Interpretation
(*With Date of Publication in* Federal Register)

Closure, Tank, and Waste Pile Standards, for Owners and Operators of Hazardous Waste
  Facilities (40 CFR Parts 264–265), *Fed. Reg.* **46,** No. 7, 2893–2897 (1981)
Hazardous Waste: Definitions of "Existing Hazardous Waste Management Facility"; Federal,
  State or Local Approvals or Permits, and Permit Prior to Construction Requirement; Interim
  Rule and Request for Comments (40 CFR Parts 122 and 260), *Fed. Reg.*, 2344–2348
  (January 9, 1981)

Source. "Regulatory Agenda, Environmental Protection Agency," 40 CFR Ch. 1, FRL #2318-1,
Fed. Reg., 18468–18517 (April 25, 1983) (issued semiannually).

---

closed or upgraded to meet the requirements of a sanitary landfill as defined by the
regulations.

Subtitle F of the RCRA requires that all federal agencies comply with all federal,
state, interstate, and local regulations stemming from the RCRA unless exempted
by the President of the United States. The RCRA must be seen in the context of
other laws. For example, it has an interface with the Safe Drinking Water Act
(SDWA) (P.L. 95-523), with the Clean Water Act (CWA) (P.L. 95-217), with the
Clean Air Act (CAA) (P.L. 95-95), and with the Hazardous Materials Transportation
Act (HMTA) (P.L. 95-403). The application of these laws to any facilities will
depend on the actual waste treatment and disposal practices. The following examples
show how the laws may overlap.

If a pathological incinerator is being used, its stack emissions must meet the
standards promulgated under CAA. Any transporting of hazardous waste will be
regulated by the HMTA, as well as by the manifest system established by the
RCRA. The sludges from water-treatment facilities must be treated as solid waste.
Some disposal methods may be prohibited or restricted; for example, landfilling or
deep-well injection could contaminate ground-water and surface-water supplies.
Open-pit burning of wastes is no longer legal (see page 81). Any treatment or
disposal facility will have to be permitted by the EPA, thus ensuring that it meets
the requirements of the environmental laws. In addition to the legislation and
regulations mentioned above, every federal agency is required by Executive Order
No. 12196 to meet their responsibilities under the Occupational Safety and Health
Act of 1970, 29 CFR Part 160.

For any facility that generates solid waste there are common problems which
must be faced by the party or parties responsible. Safety and health considerations
should include the facility personnel, contractor personnel, and the community at
large. Suitability of the treatment or disposal method must be evaluated. Energy
requirements must be a major factor in the management of a waste-treatment or
waste disposal facility. The methods of disposal must meet all federal, state, and
local regulations, and must be permitted by the U.S. EPA, state EPA, or other
appropriate state or local agencies. Other considerations include those associated
with the actual managing of a facility, such as the identification of waste and
quantity, manpower requirements, equipment limitations, and operating costs.

The EPA estimates that 10–15% of the annual production of about 344 million metric tons (wet) of industrial waste is hazardous. Quantities of hazardous waste are expected to increase by 3% annually. The EPA also estimates that 90% of hazardous waste is managed by practices that will not meet new federal standards.

Among major generators of hazardous waste, the 17 industries the EPA has studied in detail are (1977 estimates):

Organic chemicals	11.7 million metric tons (wet)
Primary metals	9.0
Electroplating	4.1
Inorganic chemicals	4.0
Textiles	1.9
Petroleum refining	1.8
Rubber and plastics	1.0
Miscellaneous (seven sectors)	1.0
Total	34.5

Seventy to 80% of these industries' hazardous waste is disposed of on the generator's property; 80% is disposed of in nonsecure ponds, lagoons, or landfills; 10% is incinerated without proper controls; and 10% is managed acceptably as compared to proposed federal standards, that is, by controlled incineration, secure landfills, and recovery. About 50% of hazardous waste is in the form of liquid or sludge.

Ten states generate 65% of all hazardous waste. They are Texas, Ohio, Pennsylvania, Louisiana, Michigan, Indiana, Illinois, Tennessee, West Virginia, and California (see Fig. 7.2).[2]

The number of hazardous-waste sites which the EPA claims constitute a health hazard in the United States has been gradually revised upward since the RCRA and SUPERFUND programs were instituted. On September 1, 1983, the EPA added 133 sites to the "national priority list" of 413 sites previously listed, all of which are now eligible for clean-up with resources of the SUPERFUND. The 40-mile stretch of the Hudson River, south of Ft. Edwards and Hudson Falls, is included as one site, since the EPA has considered this stretch contaminated with PCBs.

The total number of waste sites identified by the EPA is now 17,000, with others still under investigation. The map (see Fig. 7.3) illustrates the distribution of sites as of September 1, 1983, now having "national priority." Lee Thomas, EPA Assistant Administrator, noted that there may be as many as 22,000 abandoned chemical dumps nationwide that pose potential public-health and environmental hazards. The $1.6 billion SUPERFUND will cover only about 200 of the waste sites by some estimates. (See *New York Times,* September 2, 1983, p. A18, and the *Wall Street Journal,* September 2, 1983, p. 5. See also *EPA Annual Report,* pages 19–35, *Pollution Engineering,* Vol. XV, No. 12, December 1983.)

In another survey, conducted by the Subcommittee on Oversight and Investigations of the House Committee on Interstate and Foreign Commerce,[3] 53 of the

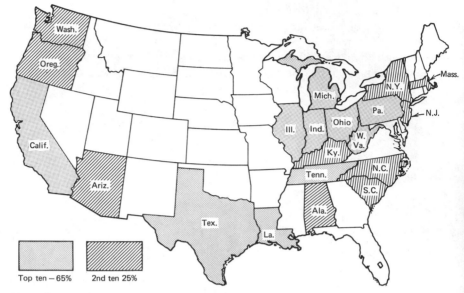

Figure 7.2. Sources of hazardous waste by states in the United States. Courtesy of *Science*, American Association for the Advancement of Science.

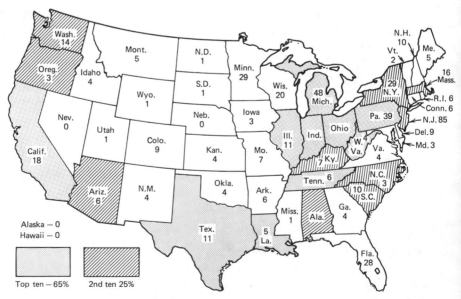

Figure 7.3. Hazardous waste sites by states: locations given priority for cleanup. See also Appendix 1, this text. Source: U.S. EPA and *The New York Times*, September 2, 1983.

largest domestic chemical companies were asked to give information on their waste generation and disposal (both nonhazardous and hazardous). This survey disclosed that approximately 66 million tons of chemical process wastes were generated in 1978 by the 1605 chemical plants of the 53 companies. No conclusion could be drawn as to what percent of all chemical wastes this amount represents, nor is it known in the absence of final federal definitions [not available at the time of the survey (1978–1979)] what percent of the 66 million tons of chemical process waste would be classified as hazardous. The EPA estimated that about 379 million tons of industrial wastes were generated in 1977 by all industry, of which the EPA then estimated approximately 39 million tons were hazardous. Thus, the 66 million tons of chemical process wastes generated in 1978 by the 1605 facilities participating in the survey relates to about 17% of the 379 million tons of industrial wastes which the EPA estimated were generated in 1977. (An adjustment was made between the two studies to reflect tons, from the metric tons used originally by the EPA study. One metric ton is 2240 lb.)

In 1982, the EPA estimated that 59 million metric tons of hazardous waste is generated annually in addition to 33 million metric tons of municipal wastes disposed of annually in 18,500 sites covering a total of 500,000 acres. Also 5 million metric tons of municipal waste water treatment sludge, and over 120 million metric tons of flue gas cleaning sludges will be generated annually by 1985, enough to cover 10 square miles with a 9-ft sludge layer.

The House Committee survey revealed that approximately 762 million tons of chemical wastes generated by the 53 participating companies since 1950, or earlier, have been disposed of in 3383 locations known to the companies. These sites do not necessarily pose threats to the public health or the environment. Of these 762 million tons, 94% were disposed of on the immediate property of the chemical plants and 6% were sent off-site for disposal (see Fig. 7.2).

The major hazards which have been observed from improper disposal of wastes have been noted by the EPA. Major routes for damage are:

1. Direct contact with toxic wastes.
2. Fire and explosions.
3. Ground-water contamination via leachate.
4. Surface-water contamination via runoff or overflow.
5. Air pollution via open burning, evaporation, and wind erosion.
6. Poison via the food chain (bioaccumulation).

The EPA has documented over 500 cases of damage to public health or the environment due to improper hazardous-waste management.[4-7]

The Department of Transportation has proposed regulations pursuant to the Hazardous Materials Transportation Act, pertaining to transportation of hazardous waste, which were published in the *Federal Register,* May 25, 1978. To the chemist or engineer, as well as to management and legal staff, the question of what is or is not a hazardous waste is paramount. Two routes or approaches determine the answer.

Certain wastes are clearly defined by official notes in the *Federal Register*, as previously noted. Other wastes are either hazardous or nonhazardous depending on whether they do or do not meet the testing criteria[8a,b] for certain characteristics. These characteristics are:

1. Ignitability
   a. For liquids, has a flash point less than 60°C (140°F) by the specified methods.
   b. Nonliquids are listed on the basis of being capable of ignition under normal conditions of spontaneous and sustained combustion.
   c. An ignitable compressed gas per Department of Transportation regulations.
   d. An oxidizer per Department of Transportation regulations. The EPA hazard code for ignitability is "I," with EPA hazardous-waste number D001.

2. Corrosivity
   a. pH is less than 2 or greater than 12.5.
   b. Corrodes steel at a rate greater than 1/4 in./year.
   c. The EPA hazard code is "C", with EPA hazardous-waste number D002.

3. Reactivity
   a. Normally unstable—reacts violently.
   b. Reacts violently with water.
   c. Forms explosive mixture with water.
   d. When mixed with water, generates toxic gases, vapors, or fumes.
   e. Contains cyanide or sulfide and generates toxic gases, vapors, or fumes at pH between 2 and 12.5.
   f. Capable of detonation if heated under confinement or subjected to strong initiating source.
   g. Capable of detonation at standard temperature and pressure.
   h. Listed by the Department of Transportation as Class A or Class B explosive.
   i. The EPA hazardous-waste number is D003.

4. EPA Toxicity
   If extract of waste contains concentrations (in mg/liter) greater than:

Arsenic	5.0
Barium	100.0
Cadmium	1.0
Chromium	5.0
Lead	5.0
Mercury	0.2
Selenium	1.0

Silver	5.0
Endrin	0.02
Lindane	0.1
Methoxychlor	10.0
Toxaphene	0.5
2,4-D	10.0
2,4,5-TP	1.0

The concentrations listed above are 100 times the concentrations listed in the Pure Drinking Water Standards. No extract is necessary if waste is liquid containing less than 0.5% filtrate solids. The EPA hazardous-waste number of each of the above constituents is D004–D017. The EPA hazard code is "E".

The criteria for listing of hazardous waste as acutely toxic or hazardous are:

a. Small doses can cause human mortality.
b. Oral LD_{50}—less than 50 mg/kg.
c. Inhalation LC_{50}—less than 2 mg/liter.
d. Dermal LD_{50}—less than 200 mg/kg.
e. Otherwise, causes or contributes to increased mortality; causes serious irreversible illness, or serious incapacitating reversible illness (see Chaps. 2 and 3).

(Radioactive and infectious wastes are treated under separate sections.)

In addition to the legal aspects of hazardous wastes, the potential economic aspects of improper disposal should be considered. Table 7.2, from the basic report prepared by the Library of Congress for the Committee on Environment and Public Works of the U.S. Senate, gives graphic illustration of the potential financial problems which have already surfaced.

Although land disposal in "approved" secure sites, where the leachate as of August 1983 is assumed not to leave the site due to proper barriers, is a relatively simple and popular approach, there are several other alternatives which might be considered. These include:

1. Storage under the proper conditions above ground, so the condition of the containers can be checked frequently and any potential problems removed before serious loss of material to the environment.
2. Landfilling, which is a form of land disposal, can be useful if the geology and other aspects are correct. However, it should be recalled that Love Canal began as a simple landfilling operation (see Fig 7.4 and Table 7.3).[9]
3. Laboratory analysis, so the exact composition of the wastes is known, is essential before any safe or legal disposal is attempted.

TABLE 7.2. Selected Cases Illustrating Financial Aspects of Improper Disposal of Wastes

State	Location	Type of Pollution (Date Discovered)	Source	Type of Injury	Compensation Sought	Compensation Obtained
Alabama	Pickwick Reservoir and impounded tributaries near Mobile	Mercury in water (1970)	Dumping	Latent: possible effects from eating fish contaminated with Hg	$157 million	Settled for undisclosed amount.
Alabama/Georgia	Lake Weiss, Coosa River	PCBs in water (1976)	Plant in Rome, Georgia	Latent: possible carcinogenic, mutagenic, and teratogenic effects from eating fish contaminated with PCB	$1.6 billion	Private plaintiffs settled for undisclosed amount. Alabama settled its $100 million suit for $67,900.
California	Lathrop	Pesticides in ground; DBCP, alpha-BHC and lindane, and radioactivity in ground water (1979)	Company plant	Acute: none reported unless worker sterility also due to groundwater pollution Latent: possible carcinogenic, mutagenic, and teratogenic effects from pesticides and radioactivity in ground water Natural resources contamination of ground water Related: three domestic wells closed, other wells under study; possible property-value decrease	$30 million in civil penalties; $15 million clean-up	None to date (June 1980).

State	Location	Event	Cause	Effects	Costs	Outcome
California	Riverside		Stringfellow site used between 1955–1972		$10 million clean-up committed by the EPA,[a] but other estimates of clean-up range to $40 million[b]	(Wall Street Journal, page 4 Dec. 12, 1983) None to date
Colorado	Rocky Mountain Arsenal, U.S. Army		pesticide mfgr.		$1.85 billion sought	
Florida[c]	Youngstown	Chlorine in air (1978)	Derailed rail tank car	Acute: 8 dead, 138 injured. Latent: possible chronic injury to respiratory tract; $1.089 million property damage; pollution of water in wells; injured persons unable to work	About $750 million	
Michigan	Montague	Toxic chlorinated hydrocarbons in ground water (1976)	Plant waste disposal	Latent: Possible carcinogenic, mutagenic, and teratogenic effects. Natural resources: 2×10^9 gal of ground-water contamination. About 12 wells contaminated. Possible property-value decline	Fine and $15 million clean-up	Settled for $15 million clean-up and $1 million fine. Performance bond posted.
Michigan	Lake Erie	Mercury in lake (1969)	Discharges from several companies	Latent: possible teratogenic effects from eating fish contaminated with Hg	About $60 million	Claimed $59 million; won $120,000

TABLE 7.2. (Continued)

State	Location	Type of Pollution (Date Discovered)	Source	Type of Injury	Compensation Sought	Compensation Obtained
Missouri	Times Beach and Minker Stout area	Dioxin	From mixing of wastes used to oil roads and horse arenas		$33 million (EPA); $3.3 million (Missouri) to buy out home owners (February, 1983)[d]	
New York State	Hudson Valley (seven sites)	46,000 tons PCB and other wastes	Dumped by contract haulers		Clean-up estimated at least $30 million[e]	
Pennsylvania	Bruin Lagoon site (Bear Creek, Allegheny River tributary)	Open lagoon with 35,000 yds of oily/acidic sludge plus 55,000 cubic yds soil	Bruin Oil Co., 1930–1970		Clean-up and disposal $24.4 million (estimate for the EPA by WESTON, 1982)	

Source: Adapted from "Six Case Studies of Compensation for Toxic Substances Pollution: Alabama, California, Michigan, Missouri, New Jersey, and Texas," prepared under the supervision of the Congressional Research Service of the Library of Congress, 96th Congress, 2d Session, Serial No. 96-13, June 1980, U.S. GPO, Washington, D.C.

[a]*Wall Street Journal,* August 3, 1983, p. 1.

[b]*Time,* April 11, 1983, p. 18.

[c](For details, see RAR-78-7, National Transportation Safety Board, Washington, D.C.)

[d]"Governments to Buy Homes in Two Towns Tainted by Dioxin," *Wall Street Journal,* June 9, 1983, p. 14.

[e]"GE Agrees to Clean Up Toxic Wastes at 7 Sites in the Hudson Valley," *Wall Street Journal,* September 25, 1980, p. 10.

TABLE 7.3.   Concentrations of Some Toxic Organic Chemicals Found in Houses Near Love Canal Dump Site

Chemical	Highest Concentration Observed[a] ($\mu g/m^3$)
Chloroform	24
Benzene	270
Trichloroethylene	73
Toluene	570
Perchloroethylene	1140
Chlorobenzene	240
Chlorotoluene	6700
Xylene (meta + para)	140
Xylene (ortho)	73
1,3,5-Trichlorobenzene	74
Total organics	12,919
(one sample, dining room)	

Source: "In the Matter of the Love Canal Chemical Waste Landfill Site," Report of the Commissioner of Health of the State of New York, August 1978; *Chem. Eng. News,* 6 (August 7, 1978).

[a]Except as noted for total organics, concentrations were measured in basements of homes.

4. Solar evaporation has limited application in most parts of the United States, but may be a useful approach where the combination of limited rainfall and long cloudless days makes the evaporation possible. One achieves some evaporation, even of aqueous wastes, when the lagoon is in the sun and wind and the humidity is low.

5. Acid neutralization is a time-tested procedure used by many industries for acid substances. Normally this involves reacting the acid with a base such as lime or limestone, forming a neutral solid. Every hazardous-waste control facility which handles acid wastes should consider this. In addition, of course, highly alkaline materials can be neutralized by acid to bring them into a nonhazardous category.

6. Chemical fixation is a solidification process by which liquid wastes are mixed with a solid material which then is disposed as a solid.

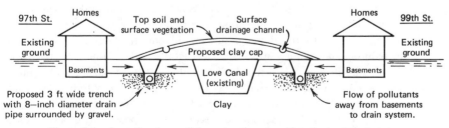

LOVE CANAL REMEDIAL CONSTRUCTION PLAN

Figure 7.4. Love Canal remedial construction plan. Note the leachate drains.

7. Treatment and detoxification involves chemical reactions to change the chemical composition to that of a more innocuous material. Oxidation is a frequently used method of change. Among materials which lend themselves to such treatment are the mercaptans, sulfides, formaldehyde, and cyanide. Oxidation is accomplished by such agents as chlorine, ozone, hydrogen peroxide, and permanganates.

8. Hydrolysis is the process by which a reaction with moisture causes the change in composition by evolution of a toxic gas, such as HCl or HBr. Organophosphate esters lend themselves to hydrolysis by strong alkali.

9. Deep-well injection is a disposal method which has been widely used in those areas where the material can be filtered, and then injected under pressure into a wellhead into a permeable zone deep beneath the aquifer, and then into an impermeable zone. A permeable zone may be sandstone or sand. An impenetrable zone is rock or other strata between the aquifer, sealing it off. The main objection to this process is that the material may eventually leak, causing contamination in the water supplies or stream.

10. Farming is a method of disposal in which certain materials can be spread on land, or crops, permitting soil microorganisms to attack it, hoping that what is left is a fertilizer. Much attention must be given to heavy metals and to chlorinated hydrocarbon concentrations.[9b] The method is losing favor at present.

11. Dewatering is an important step, especially if the material is to be evaporated or transported. Filtration by diatomous earth or activated carbon is often used.

12. Incineration under proper conditions is the most attractive method of disposal if the material has no further value. Even highly halogenated materials can be disposed of in this manner, if the proper temperatures, residence time, turbulence, and afterburners, with appropriate scrubbers, are installed and maintained properly.[9c,d] The residue or "ashes" from incinerators must, of course, be disposed of, usually by approved landfilling (Fig. 7.4). If landfilling or burial is undertaken, the proper procedures, including a leachate collection system, should be followed.

13. Recycling, which is the process of reusing wastes, has become of much interest, since with treatment, materials may often be used more than once in the cycle, or by another organization with completely different needs. As an example, #5 fuel oil may be reclaimed from hazardous waste and burned as fuel. Automobile crankcase drainings, with proper treatment, are also highly useful after reclaimed. Solvents from processes as varied as painting and printing are being successfully and economically recycled. One authority has stated that between 50 to 80% of hazardous chemical wastes can be recycled, with economic as well as environmental gain. Figures 7.5–7.7 suggest how information about waste change can be of benefit. There has been serious discussions of the possibility of a Recycling Institute, which would make technical information available to encourage

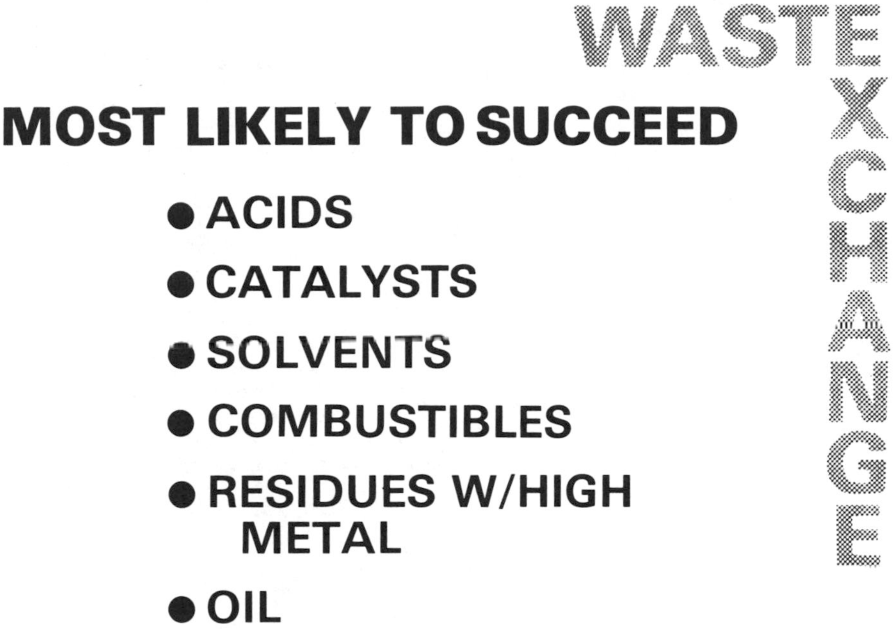

WASTE
EXCHANGE
MOST LIKELY TO SUCCEED
• ACIDS
• CATALYSTS
• SOLVENTS
• COMBUSTIBLES
• RESIDUES W/HIGH METAL
• OIL

Figure 7.5. Waste exchange: substances most likely to be exchanged. Courtesy of U.S. EPA.

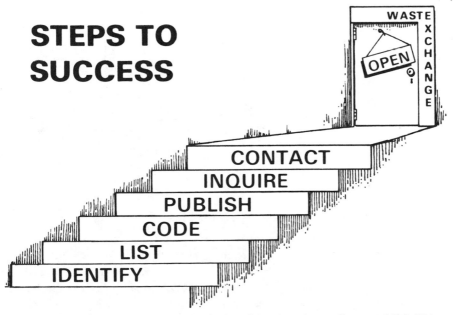

STEPS TO SUCCESS
WASTE EXCHANGE
OPEN
CONTACT
INQUIRE
PUBLISH
CODE
LIST
IDENTIFY

Figure 7.6. Steps to organization and operation of a waste exchange. Courtesy of U.S. EPA.

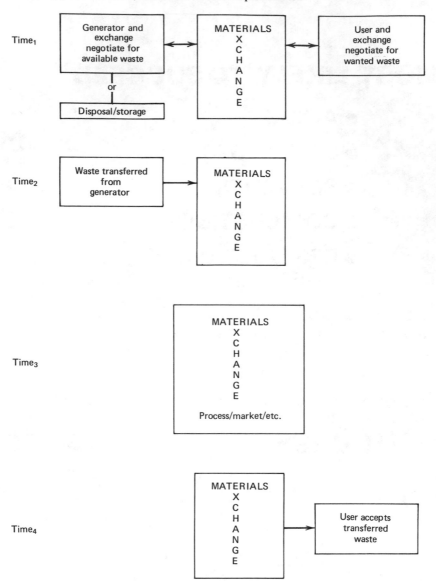

Figure 7.7. Sequence of organization and operation of a waste exchange. Courtesy of U.S. EPA.

the recycling of many wastes which currently must be disposed of. The Midwest Industrial Waste Exchange, 10 Broadway, St. Louis, MO 63102, is an organization which lists availability of materials, and directs the potential user to a source. An annual catalog of wastes is issued to members. Another group is the Pensylvania Waste Exchange, which is a service of the Pennsylvania Chamber of Commerce, Harrisburg, Pennsylvania (Table 7.4).

Recognizing that the RCRA regulations will present many problems in compliance, the U.S. EPA has established a toll-free service to provide assistance to industry and interested citizens. The number is (800)424-9346. Those in the Washington, D.C. area should use (202)554-1404. Calls will be answered by trained professionals Monday through Friday from 9 a.m. to 4:30 p.m. EST. By mail, assistance may be obtained by writing to the Industry Assistance Office, EPA Office of Solid Waste (WH-565), Washington, D.C. 20460.

It should be noted this is the second toll-free "hot line" which has been established by the EPA to extend assitance to industry. The first, for inquirers about the Toxic Substances Control Act, is (800)424-9065. [In the Washington, D.C. area, call (202)554-1404.] Mail inquiries regarding the TSCA should be directed to the Industry Assistance Office, Office of Toxic Substances (TS-799), U.S. Environmental Protection Agency, Washington, D.C. 20460. In addition to answering specific inquiries, this office maintains an extensive mailing list, so documents and other regulatory reviews may be mailed on a routine basis.

TABLE 7.4.   United States Waste Exchanges, 1983

California

Department of Health Services
   Hazardous Materials Management Section
2151 Berkeley Way
Berkeley, CA 94704
Zero Waste Systems, Inc.
2928 Poplar Street
Oakland, CA 94608

Connecticut

World Association for Safe Transfer and
   Exchange (WASTE)
130 Freight Street
Waterbury, CN 06702

Georgia

Georgia Waste Exchange
   Georgia Business & Industry
   Association
181 Washington St. SW
Atlanta, GA 30303

Illinois

Environmental Clearinghouse
   Organization—ECHO
3426 Maple Lane
Hazel Crest, IL 60429

American Chemical Exchange (ACE)
   4849 Golf Rd.
   Skokie, IL 60076

Indiana

Waste Materials Clearinghouse,
   Environmental Quality Control, Inc.
1220 Waterway Boulevard
Indianapolis, IN 46202

Iowa

Iowa Industrial Waste Information Exchange
   Center for Industrial Research & Service
   201 Building E
   Iowa State University
   Ames, IA 50011

Massachusetts

The Exchange
   63 Rutland Street
   Boston, MA 02118

Michigan

American Materials Exchange Network
   19489 Lahser Road
   Detroit, MI 48219

TABLE 7.4. *(Continued)*

Minnesota

Minnesota Association of Commerce &
Industry (MACI)
200 Hanover Building
480 Cedar Street
St. Paul, MN 55101

Missouri

Midwest Industrial Waste Exchange
920 Main Street
Kansas City, MO 64105

New Jersey

Industrial Waste Information Exchange
New Jersey State Chamber of Commerce
5 Commerce Street
Newark, NJ 07102

New York

Enkarn Research Corporation
P.O. Box 590
Albany, NY 12201
The American Alliance of Resources
Recovery Interests, Inc. (AARRII)
111 Washington Avenue
Albany, NY 12210

North Carolina

Mecklenburg County Waste Exchange
Mecklenburg County Engineering
Department
1501, I-85 North
Charlotte, NC 28216

Ohio

ORE Corporation—"The Ohio
Resource Exchange"
Columbus Industrial Association
1646 West Lane Avenue
Columbus, OH 43221
Industrial Waste Information Exchange
Columbus Industrial Association
1646 West Lane Avenue
Columbus, OH 43221

Oregon

Oregon Industrial Waste Information
Exchange
Western Environmental Trade Association
333 SW 5th Suite 618
Portland, OR 97204
or
Resource Conservation Consultants
1615 NW 23 Suite One
Portland, OR 97204

Pennsylvania

Pennsylvania Waste Information Exchange
Pennsylvania Waste Information
Exchange
(A service of the Pennsylvania
Chamber of Commerce)
222 North Third Street
Harrisburg, PA 17101

Tennessee

Tennessee Waste Swap
Tennessee Manufacturers Association
708 Fidelity Federal Building
Nashville, TN 37219

Texas

Chemical Recycle Information Program
Houston Chamber of Commerce
1100 Milam Building, 25th Floor
Houston, TX 77002

Washington

Information Center for Waste Exchange
2112 Third Avenue, Suite 303
Seattle, WA 98121

West Virginia

Union Carbide Corporation
Invest Recovery Department
P.O. Box 8361
Building 3005
South Charleston, WV 25303

Source: "Waste Exchanges, Background Information," SW-887-1, December 1980, Office of Water
and Waste Management, U.S. Environmental Protection Agency, Washington, D.C.

"SUPERFUND" and Its Implications for the Chemist, Manager, and Engineer

With the passage of P.L. 96-510, the Comprehensive Environmental Response, Compensation, and Liability Act of 1980, signed by President Carter on December 11, 1980, a 5-year, $1.6 billion trust was established to clean up hazardous chemical spills and abandoned waste dumps, and to respond, with remedial action, to the threat of hazardous-material release. According to the bill, industry fees contribute 87.5% or $1.38 billion and appropriations of general revenues supply $220 million. Fees are collected on 42 specified substances: 65% of industry's share comes from taxes on petrochemicals, 20% on inorganic chemicals, and 15% on crude oil. Liability is equivalent to that in Section 311 of the Clean Water Act, which courts have interpreted as strict liability. Common law has established joint and multiple liability for hazardous-waste activities with the liable party responsible for costs of removal or remedial action incurred by the government, and for restoring lost natural resources up to $50 million. Operators of hazardous-waste disposal sites will be taxed to set up a $200 million postclosure fund to cover the costs of monitoring and maintaining disposal sites after closure. An agency for a toxic substances and disease registry will be set up within the Public Health Service. Failure to notify the proper agency of the existence of a disposal facility can elicit fines of up to $10,000 and/or up to 1 year in jail. Failure to notify the agency of a release can incur fines of up to three times the cost of clean-up.

With this amount of money, and the public demand for prompt action, it is inevitable that the chemist and engineer will be called upon to analyze hazardous waste itself (relatively easy if a properly equipped laboratory and trained personnel are available to carry out the test procedures outlined by the implementation of the RCRA act), and to pass judgment on the questions which doubtlessly will arise from past practices of waste disposal. While some final regulations and criteria have not been issued for disposal-site evaluations, and doubtlessly will vary from state to state insofar as implementation is concerned, fundamental knowledge can be applied in a systematic method to more clearly elucidate the problems and their potential solutions. In the final report "Methodology for Rating the Hazard Potential of Waste Disposal Sites," May 5, 1980, prepared for the EPA Office of Research and Development by JRB Associates, several aids are offered to the rating and evaluation of sites insofar as waste disposal is concerned.[10] Included are a worksheet for waste-disposal sites and a rating form. This material is quoted in Tables 7.5–7.7. Included is material on the relative persistence or biodegradability of certain organic compounds, quoted from the report by E. F. Abrams. The reader may also find the book *Biochemical Toxicology of Environmental Agents* by A. DeBruin, Elsevier, Amsterdam, 1976 to be useful.[11]

One of the more interesting aspects of waste disposal, which brings into focus a largely overlooked field, is that of waste compatibility, or incompatibility. The subject has developed in importance over the years, as more and more chemical combinations are possible, and chemists are apparently not always sufficiently aware of the consequences. Fawcett published a preliminary list of then recognized incompatible substances in 1952.[12] The National Fire Protection Association recognized the problem in 1953, and the author organized and chaired the committee which reviewed and expanded on the work of the late George Jones. From this resulted the NFPA 491-M *Manual of Hazardous Chemical Reactions,* which has undergone five revisions and lists over 2000 reactions known to be hazardous.[13] The excellent exploratory work which the Dow Chemical Company did under contract with the National Research Council, in behalf of the Committee on Hazardous Materials, provided guidance to the U.S. Coast Guard in regulating the storage and handling of hazardous cargoes on barges and ships.[14] The publications of two editions of the book by Leslie Bretherick, of BP Research Laboratory, *Handbook of Reactive Chemical Hazards,* was a major improvement in available data base.[15] All the above assumed that relatively pure Chemical A would react or not react with Chemical B under certain conditions, often with spectacular results, as pictured in the motion picture *Chemical Boobytraps.*[16] However, extensive as our knowledge base is, it is relatively poor insofar as information about the reactions of chemicals in the complex matrices of wastes. Many factors greatly influence waste-component reactions. Among these are temperature, catalytic effects of dis-

Figure 7.8. ''Valley of the Drums'' illustrates the economic, as well as human and environmental impact of improper hazardous waste disposal. Courtesy of the Louisville, Kentucky Courier-Times.

solved or particulate metals, soil reactions, and reactions with surfaces to transport vehicles or containers. For these reasons, a study of the problem by the California Department of Health Services for the Solid and Hazardous Waste Research Division of the Municipal Environmental Research Laboratory should be of great interest, and the essence of that study is included from their final report[17] (Table 7.A.3). If the above assignments seem unproductive, they should be viewed in terms of risk/ benefit, human health, and environmental impact which may be significant over many years, to say nothing of the potential legal and financial liability and public-relation aspects to the generator, transporter, and disposal-site owner. The alternative is to reengineer the process or operation so hazardous waste is not a byproduct, or part of a closed system. *It is the responsibility of chemists, managers, and engineers to ensure that such illicit dumping of hazardous wastes, as shown in Figs. 7.1, 7.8, and 8.6 will not reoccur.*

If complete knowledge of wastes is not available, sampling and analysis may be necessary (see Fig. 7.9).

Figure 7.9. Sampling and analyses may be required to establish essential data for waste-site management, such as at Love Canal and Times Beach. Dioxin and other substances were found in sewer and drainage systems. Courtesy of *Geomet Technology*.

TABLE 7.5. Rating Factors and Scales for Each of the Four Generic Areas (Receptors, Pathways, Waste Characteristics, and Waste-Management Practices)

Rating Factors	Rating Scale Levels			
	0	1	2	3
Receptors				
Population within 1000 ft	0	1–25	26–100	Greater than 100
Distance to nearest drinking water well	Greater than 3 miles	1–3 miles	3001 ft–1 mile	0–3000 ft
Distance to nearest off-side building	Greater than 2 miles	1–2 miles	1001 ft–1 mile	0–1000 ft
Land use/zoning	Completely remote (zoning not applicable)	Agricultural	Commercial or industrial	Residential
Critical environments	Not a critical environment	Pristine natural areas	Wetlands, flood plains, and preserved areas	Major habitat of an endangered or threatened species
Pathways				
Evidence of contamination	No contamination	Indirect evidence	Positive proof from direct observation	Positive proof from laboratory analyses
Level of contamination	No contamination	Low levels, trace levels, or unknown levels	Moderate levels or levels that cannot be sensed during a site visit but which can be confirmed by a laboratory analysis	High levels or levels that can be sensed easily by investigators during a site visit
Type of contamination	No contamination	Soil contamination only	Biota contamination	Air, water, or foodstuff contamination
Distance to nearest surface water	Greater than 5 miles	1–5 miles	1001 ft–1 mile	0–1000 ft
Depth to ground water	Greater than 100 ft	51–100 ft	21–50 ft	0–20 ft
Net precipitation	Less than −10 in.	−10 to +5 in.	+5 to +20 in.	Greater than +20 in.
Soil permeability	Greater than 50% clay	30–50% clay	15–30% clay	0–15% clay
Bedrock permeability	Impermeable	Relatively impermeable	Relatively permeable	Very permeable
Depth to bedrock	Greater than 60 ft	31–60 ft	11–30 ft	0–10 ft

Waste Characteristics

Toxicity	Sax's level 0 or NFPA's level 0	Sax's level 1 or NFPA's level 1	Sax's level 2 or NFPA's level 2	Sax's level 3 or NFPA's levels 3 or 4
Radioactivity	At or below background levels	1–3 times background levels	3–5 times background levels	Over 5 times background levels
Persistence	Easily biodegradable compounds	Straight chain hydrocarbons	Substituted and other ring compounds	Metals, polycyclic compounds, and halogenated hydrocarbons
Ignitability	Flash point greater than 200°F or NFPA's level 0	Flash point of 140°–200°F, or NFPA's level 1	Flash point of 80°–140°F or NFPA's level 2	Flash point less than 80°F, or NFPA's levels 3 or 4
Reactivity	NFPA's level 0	NFPA's level 1	NFPA's level 2	NFPA's levels 3 or 4
Corrosiveness	pH of 6–9	pH of 5–6 or 9–10	pH of 3–5 or 10–12	pH of 1–3 or 12–14
Solubility	Insoluble	Slightly soluble	Soluble	Very soluble
Volatility	Vapor pressure less than 0.1 mm Hg	Vapor pressure of 0.1–25 mm Hg	Vapor pressure of 25–78 mm Hg	Vapor pressure greater than 78 mm Hg
Physical state	Solid	Sludge	Liquid	Gas

Waste-Management Practices

Site security	Secure fence with lock	Security guard but no fence	Remote location or breachable fence	No barriers
Hazardous-waste quantity	0–250 tons	251–1000 tons	1001–2000 tons	Greater than 2000 tons
Total waste quantity	0–10 acre ft	11–100 acre ft	101–250 acre ft	Greater than 250 acre ft
Waste incompatibility	No incompatible wastes are present	Present, but does not pose a hazard	Present and may pose a future hazard	Present and posing an immediate hazard
Use of liners	Clay or other liner resistant to organic compounds	Synthetic or concrete liner	Asphalt-base liner	No liner used
Use of leachate collection systems	Adequate collection and treatment	Inadequate collection or treatment	Inadequate collection and treatment	No collection or treatment
Use of gas collection systems	Adequate collection and treatment	Collection and controlled flaring	Venting or inadequate treatment	No collection or treatment
Use and condition of containers	Containers are used and appear to be in good condition	Containers are used but a few are leaking	Containers are used but many are leaking	No containers are used

Information Requirements

Data availability is the key to the usefulness of a rating system for waste-disposal sites. Experience has shown that systems that rely on generally available information are the most widely used. In addition, systems that can tolerate data gaps are used more commonly than those systems that cannot. As a result, this rating methodology was designed to be used primarily with readily available information. Sources of this information are identified in Section 7.1. The methodology has also been made flexible enough to allow sites to be rated even if some data are not available. Section 7.2 describes two methods for resolving missing-data problems when rating a site.

7.1. DATA SOURCES

The first consideration in designing the rating methodology would be to determine who would be using the system, and where would the necessary information be found. As a result of this consideration, we have identified and compiled data sources for each rating factor. Table 7.6 presents a list of the 31 rating factors together with possible sources of information for each. The various sources of data can be grouped into four categories:

1. Standard references and indices.
2. Other published sources.
3. Contacts with knowledgeable parties.
4. Site visits.

The first category includes Sax's *Hazardous Properties of Industrial Materials,* the National Fire Protection Association's *Guide to the Hazardous Properties of Chemicals,* CRC's *Handbook of Chemistry and Physics,* Lang's *Handbook of Chemistry,* and similar publications. The second category includes the *Federal Register* and technical journals, maps, and reports from the U.S. Geological Survey (USGS), the National Oceanographic and Atmospheric Administration (NOAA), and the Soil Conservation Service. Also included are environmental impact statements for projects in the area of the site and other publications from environmental groups, such as the National Wildlife Federation, and state and local governments. The third category (via telephone, mail, or in-person contacts) includes individuals working in industry (e.g., waste generators, transporters, or managers): federal, state, and local government employees; or persons involved in environmental action groups as well as some nearby-site residents. The fourth and last category refers to cases where the rater or some other trained investigator actually visited the site in question.

Additional sources of information which may be useful in the appraisal of chemical and toxic substances for handling and disposal were identified at the 186th

National Meeting of the American Chemical Society, which took place in Washington, D.C. on August 29, 1983 to September 2, 1983. These included:

The electronic media, such as the local and national television networks, the radio stations (both FM and AM), and video tapes.

The print media, including magazines, newspapers, and newsletters, which often cover stories in depth where the interest warrants it. An example is the *Wall Street Journal* coverage of the dioxin sampling program at an abandoned chemical plant in Newark, New Jersey (R. Winslow, "Hunting for Dioxin Exposes Field Teams to Array of Hazards," *Wall Street Journal,* July 26, 1983, p. 1.).

The American Chemical Society Health and Safety Referral Service, located at the society's headquarters library in Washington, D.C., which is frequently the first step in a referral system. The service may be contacted by telephoning (202) 872-4511.

The Library of Congress National Referral Center, which maintains an automated data bank of 13,000 information resources, that is, organizations that collect and disseminate specialized information. Many of these organizations can be useful to the chemist and engineer. The Library of Congress can be contacted by telephoning (202) 287-5671 or writing to the National Referral Center, Washington, D.C. 20540.

The National Library of Medicine, which is a part of the National Institutes of Health complex and maintains extensive chemical and toxicological files in their specialized information services. Through the MEDLARS system, several databases are available which are of significant value in the description and control of chemical hazards. These specialized computer services include TOXLINE, CHEMLINE, RTECS, and TDB. All are available to the public under the management of the Toxicology Information Program of the library. The Toxicology Data Bank (TDB) is a peer-reviewed, fact-retrieval file containing comprehensive toxicity and chemical information; mutagenicity, teratogenicity, and carcinogenicity data; enviromental and occupational information; and the procedures for antidotes and treatments. The library is located in Bethesda, Maryland, and the telephone number is (301) 496-1131.

The National Academy of Sciences/National Research Council/National Academy of Engineering/Institute of Medicine, which maintains a Toxicology Information Center (TIC) that provides information on the toxicity of chemicals to humans, including industrial chemicals, environmental pollutants, and commercial products. Information includes animal studies, human studies, and studies on the interpretation of toxicity data (such as risk assessment, structure-activity relationships, and extrapolation). The TIC's collection of over 100,000 documents includes journals, monographs, technical reports, government documents, and unpublished materials. The collection dates back to the nineteenth century and is extensive for the years before 1966. The collection is indexed from the viewpoint of the practicing toxicologist in a card catalog which is presently being automated. The TIC supports committees under the Board on

TABLE 7.6. Sources of Information for Each of the Rating Factors

Rating Factor	Sources of Information
Receptors	
Population within 1000 ft	Local housing officials or census officers
	Current topographic maps or aerial photos
Distance to nearest drinking water well	Information obtained from knowledgeable sources such as the Public Health Departments, water-supply companies, well drillers, residents
Distance to nearest off-site building	Local housing officials or census officers
	Current topographic maps or aerial photos
Land use/zoning	Land use or zoning maps
	Aerial photos (see Fig. 7.10)
Critical environments	National Wildlife Federation and other national environmental groups
	State and local environmental groups
	U.S. Fish and Wildlife Service
	State departments of fish and game
Pathways	
Evidence of contamination	Information obtained from knowledgeable parties
Level of contamination	Information obtained from knowledgeable parties
Type of contamination	Information obtained from knowledgeable parties
Distance to nearest surface water	USGS topographic maps or reports
	Maps and reports from state or local highway departments or from universities or state geological surveys (see Fig. 7.11)
Depth to ground water	USGS water-supply papers, ground-water bulletins, and geologic reports
	Local well drillers, water suppliers, and universities (geology departments)
Net precipitation	NOAA annual weather summaries
	General precipitation and evapotranspiration maps
Soil permeability	USDA Soil Conservation Service county maps and reports
	USGS soil maps and reports (see Fig. 7.12)
Bedrock permeability	USGS water-supply papers, ground-water bulletins, and geologic reports
	Local well drillers, water suppliers, and universities (geology departments)
Depth to bedrock	USDA Soil Conservation Service county maps and reports
	USGS soil and geologic maps and reports
Waste Characteristics	
Toxicity	*Hazardous Properties of Industrial Materials* by N. I. Sax
	NFPA's *Guide on Hazardous Materials*
	Registry of Toxic Effects of Chemical Substances
Radioactivity	Information obtained from knowledgeable parties; survey

TABLE 7.6.   (*Continued*)

Rating Factor	Sources of Information
Persistence	Appendix A Table 7.A.2.
	Partition coefficients [see "Partition Coefficients and Bioaccumulation of Selected Organic Chemicals," *Env. Science Tech.* **II,** No. 5, 475 (May 1977)]
Ignitability	NFPA's *Guide on Hazardous Materials*
	Lang's *Handbook of Chemistry*
Reactivity	NFPA's *Guide on Hazardous Materials;* Bretherick's book
	Proposed RCRA Regulations, *Fed. Reg.* (December 18, 1978)
Corrosiveness	Information obtained from knowledgeable parties; analysis (see Fig. 7.9)
Solubility	CRC's *Handbook of Chemistry and Physics*
	Lang's *Handbook of Chemistry*
	Merck Index
	Handbook of Environmental Data on Organic Chemicals
Volatility	CRC's *Handbook of Chemistry and Physics*
	Lang's *Handbook of Chemistry*
	Handbook of Environmental Data on Organic Chemicals
Physical state	Information obtained from knowledgeable parties
Waste-Management Practices	
Site security	Information obtained from knowledgeable parties
Hazardous-waste quantity	Information obtained from knowledgeable parties
Total waste quantity	Information obtained from knowledgeable parties
Waste incompatibility	Appendix A
Use of liners	Information obtained from knowledgeable parties
Use of leachate collection systems	Information obtained from knowledgeable parties
Use of gas collection systems	Information obtained from knowledgeable parties
Use and condition of containers	Information obtained from knowledgeable parties

Toxicology and Environmental Health Hazards of the NRC. The telephone number in Washington, D.C. is (202) 334-2000.

The International Occupational Safety and Health Information Center (CIS), which is a unit of the International Labor Office (part of the United Nations). *This is not a labor union or labor oriented.* Since 1960 the CIS has published abstracts and indexes of the most important literature covering all aspects of safety and health in the workplace, which has direct application to chemical and toxic-chemical handling and disposal. The literature comes from worldwide sources and is contributed, in part, by a network of CIS National Centers in 40 countries. The U.S. National Center for CIS is the NIOSH Clearinghouse for Occupational Safety and Health. On-line access to the CIS database is now

Figure 7.10. Aerial investigation of waste site. Three tanker trucks discharge liquid into waste ponds between refinery and residential area. Subsequent photos revealed entire pit area was filled in. Courtesy of U.S. EPA, Las Vegas, Nevada Office.

available from QUESTEL. In the United States write to the NIOSH Clearinghouse for Occupational Safety and Health Information, 4676 Columbia Parkway, Cincinnati, OH 45226. The International Center is located in Geneva, Switzerland.

The National Safety Council is a nonprofit organization founded in 1915, and is chartered by the Congress of the United States to coordinate safety in the broad sense. Several sections in the Industrial Department of the Council, such as the Chemical and R&D Sections, as well as the reference library, are available as information sources. Write to the National Safety Council, 444 North Michigan Avenue, Chicago, IL 60611 or telephone (312) 527-4800.

The National Fire Protection Association (NFPA), founded in 1896 as a clearinghouse for information among fire-protection and insurance specialists, which is actively involved in standards writing and providing technical services for its membership and the public in areas of concern relating to fire protection. In the

Figure 7.11. Outdoor storage of hazardous wastes in quantities above authorized amounts resulted in major fine against plant. Note proximity to major waterway (river) in foreground. Courtesy of Schenectady, New York *Gazette*, by S. M. Brown. Suggested by G. H. Proper, Jr.

chemical- and toxic-waste area, several standards and manuals are of direct applications, such as NFPA 49, 491-M, 325-M, and 704-M. For full information, contact the Association at Batterymarch Square, Quincy, MA 02269.

On-line structure/property correlations of self-reactive chemical hazards, which have been studied using the Advanced Wiswesser Line Notation (AWLN). Contact Mr. W. J. Wiswesser at (301) 663-7132 in Frederick, Maryland for information.

The Society of Toxicology (SOT), which is a national professional society devoted to the science of toxicology. The national headquarters are located at 475 Wolf Ledges Parkway, Akron, OH 44311 and the telephone number is (216) 762-7294.

The Compressed Gas Association (CGA), which is a nonprofit organization founded in 1913 and dedicated to the promotion of safety throughout the compressed-gas industry. Information on the safe handling and disposal of compressed gases is available from the CGA national office located at 1235 Jefferson Davis Highway, Arlington, VA 22202, or by telephoning (703) 979-0900.

The Safety Equipment Institute (SEI), which was founded by the Industrial Safety Equipment Association, which felt it desirable to establish a third-party certification for personal protective equipment. The SEI issues a certification mark

Figure 7.12.   Soil permeability and contamination may require soil sampling. TCE and other chemicals allegedly contaminated wells in Pennfield, Michigan Township, leading to soil sampling before extending of city water mains to serve residences. Michigan Department of Natural Resources personnel dig core samples. Courtesy of Battle Creek, Michigan *Enquirer* and Ms. Jean Miller.

to manufacturers of safety equipment. The Institute is located at 1901 North Moore Street, Suite 501, Arlington, VA 22209. The telephone number is (703) 525-1695.

The National Bureau of Standards, which conducts research and serves as a clearinghouse for information, such as on flammability and toxicity. Contact The Director, Center for Fire Research, National Bureau of Standards, Washington, D.C. 20234 or call (301) 921-3143.

The American Institute of Chemical Engineers, which established a Safety and Health Division as a service to its membership within the chemical process industries. Further information about the division may be obtained from the AIChE headquarters located at 345 East 47th Street, New York, NY 10017.

The insurance community, consisting of several interests including fire, liability, and compensation, which is a very valuable information source. The activity is coordinated by the Engineering and Safety Service, American Insurance Association, 85 John Street, New York, NY 10038. The telephone number is (212) 669-0479.

The Association of State and Territorial Solid Waste Management Officials, which coordinates the RCRA and SUPERFUND efforts at the state and local level. The address is 444 North Capitol Street, Suite 331, Washington, D.C. 20001. The telephone number is (202) 624-5828.

The Association of State and Interstate Water Pollution Control Administrators, which coordinates water pollution as interfaced by federal laws. The address is 444 North Capitol Street, Washington, D.C. 20001. The telephone is (202) 624-7782.

Earthwatch, a field research corps, which is a national, nonprofit volunteer association dedicated to encouraging and financing field research in the physical sciences. Earth, life, and marine sciences are involved in studies conducted in 18 states and 30 countries. The "on-site" observations of this organization may be very useful in ecology and solid-waste management. Write Earthwatch at 10 Juniper Road, Box 127, Belmont, MA 02178, or call (617) 489-3030.

The National Solid Waste Management Association, headquartered at 1120 Connecticut Avenue, NW, Washington, D.C. 20036 [telephone (202) 659-4513], which issues publications including a monthly magazine on waste control. It also serves as a clearinghouse for information, and conducts seminars and other educational activities.

The Chemical Manufacturers Association, located at 2501 M Street, NW, Washington, D.C. 20037 [telephone (202) 887-1100], which coordinates safety and health activities of member companies.

The Chemical Specialties Manufacturers Association, which is located at 1001 Connecticut Avenue, NW, Washington, D.C. 20036 [telephone (202) 872-8110].

7.2. DATA GAPS

Most of the information required for implementing this rating methodology will be available from the readily available data sources listed in Table 7.6. The bulk of this information should be obtained as part of an office-based data-gathering effort performed prior to a site visit. The methodology has been designed so that all factors should be rated by the time the site visit is completed. However, to develop a preliminary rating (in order to determine the preferential order of site visits) it is sometimes necessary to handle rating factors for which no data are available. There are two ways to approach this problem. First, raters can use their best technical judgment. For example, if there was no information available on a waste's physical state, a rater could either make a considered judgment based on the constituents in the waste or assume that the waste was a liquid (i.e., make a conservative, yet reasonable, assumption). Second, raters can delete factors for which there are no

TABLE 7.7. Worksheet for Rating Disposal Sites

WORK SHEET FOR RATING DISPOSAL SITES

NAME OF SITE _____ ACTIVE: INACTIVE: INACTIVE AND ABANDONED (CIRCLE ONE)

LOCATION_____

OWNER/OPERATOR _____

COMMENTS: _____

_____

_____

_____

PREPARED BY: _____ ON _____ 19 ____

FACTOR	OBSERVATION
RECEPTORS	
POPULATION WITHIN 1,000 FEET	
DISTANCE TO NEAREST DRINKING-WATER WELL	
DISTANCE TO NEAREST OFF-SITE BUILDING	
LAND USE/ZONING	
CRITICAL ENVIRONMENT	
USE OF SITE BY RESIDENTS	
USE OF NEAREST BUILDINGS	
PRESENCE OF PUBLIC WATER SUPPLIES	
PRESENCE OF AQUIFER RECHARGE AREA	
PRESENCE OF TRANS-PORTATION ROUTES	
PRESENCE OF IMPORTANT NATURAL RESOURCES	
OTHER:	
PATHWAYS	
EVIDENCE OF CONTAMINATION	
TYPE OF CONTAMINATION	
LEVEL OF CONTAMINATION	
DISTANCE TO NEAREST SURFACE WATER	
DEPTH TO GROUND WATER	
NET PRECIPITATION	
SOIL PERMEABILITY	
BEDROCK PERMEABILITY	
DEPTH TO BEDROCK	
EROSION AND RUNOFF PROBLEMS	
SUSCEPTIBILITY TO FLOODING	
SLOPE INSTABILITY	
SEISMIC ACTIVITY	
OTHER:	

180

TABLE 7.7. *(Continued)*

RATING FORM FOR WASTE DISPOSAL SITES

NAME OF SITE _____ ACTIVE: INACTIVE: INACTIVE AND ABANDONED (CIRCLE ONE)

LOCATION _____

OWNER/OPERATOR _____

COMMENTS: _____

_____

_____

_____

PREPARED BY: _____ ON _____ 19 ____

RATING FACTOR	SOURCE AND BASIS OF INFORMATION	SITE RATING (CIRCLE ONE)				MULTI-PLIER	SITE SCORE	MAXIMUM POSSIBLE SCORE
RECEPTORS								
POPULATION WITHIN 1,000 FEET		0	1	2	3	12		36
DISTANCE TO NEAREST DRINKING-WATER WELL		0	1	2	3	8		24
DISTANCE TO NEAREST OFF-SITE BUILDING		0	1	2	3	8		24
LAND USE/ZONING		0	1	2	3	6		18
CRITICAL ENVIRONMENTS		0	1	2	3	6		18
ADDITIONAL POINTS FOR OTHER RECEPTORS								50

NUMBER OF MISSING AND ASSUMED VALUES = _____ OUT OF 5.   SUBTOTALS

PERCENTAGE OF MISSING AND ASSUMED VALUES = _____ %.   SUBSCORE (SITE SCORE DIVIDED BY MAXIMUM SCORE AND MULTIPLIED BY 100.)

RATING FACTOR	SOURCE AND BASIS OF INFORMATION	SITE RATING (CIRCLE ONE)				MULTI-PLIER	SITE SCORE	MAXIMUM POSSIBLE SCORE
PATHWAYS								
EVIDENCE OF CONTAMINATION		0	1	2	3	2		6
LEVEL OF CONTAMINATION		0	1	2	3	7		21
TYPE OF CONTAMINATION		0	1	2	3	5		15
DISTANCE TO NEAREST SURFACE WATER		0	1	2	3	8		24
DEPTH TO GROUNDWATER		0	1	2	3	7		21
NET PRECIPITATION		0	1	2	3	6		18
SOIL PERMEABILITY		0	1	2	3	6		18
BEDROCK PERMEABILITY		0	1	2	3	4		12
DEPTH TO BEDROCK		0	1	2	3	4		12
ADDITIONAL POINTS FOR OTHER PATHWAYS								25

NUMBER OF MISSING AND ASSUMED VALUES = _____ OUT OF 9.   SUBTOTALS

PERCENTAGE OF MISSING AND ASSUMED VALUES = _____ %.   SUBSCORE (SITE SCORE DIVIDED BY MAXIMUM SCORE AND MULTIPLIED BY 100)

TABLE 7.7 (*Continued*)

WASTE CHARACTERISTICS	
TOXICITY	
PERSISTENCE	
RADIOACTIVITY	
IGNITABILITY	
REACTIVITY	
CORROSIVENESS	
SOLUBILITY	
VOLATILITY	
PHYSICAL STATE	
INFECTIOUSNESS	
BIOACCUMULATION POTENTIAL	
CARCINOGENICITY, TERATO-GENICITY, AND MUTAGENICITY	
OTHER:	
WASTE MANAGEMENT PRACTICES	
SITE SECURITY	
HAZARDOUS WASTE QUANTITY	
TOTAL WASTE QUANTITY	
WASTE INCOMPATIBILITY	
USE OF LINERS	
USE OF LEACHATE COLLECTION SYSTEMS	
USE OF GAS COLLECTION SYSTEMS	
USE AND CONDITION OF CONTAINERS	
LACK OF SAFETY MEASURES	
EVIDENCE OF OPEN BURNING	
DANGEROUS HEAT SOURCES	
INADEQUATE WASTE RECORDS	
INADEQUATE COVER	
OTHER:	

TABLE 7.7 (*Continued*)

NAME OF SITE _____

WASTE CHARACTERISTICS								
TOXICITY		0	1	2	3	7		21
RADIOACTIVITY		0	1	2	3	7		21
PERSISTENCE		0	1	2	3	5		15
IGNITABILITY		0	1	2	3	3		9
REACTIVITY		0	1	2	3	3		9
CORROSIVENESS		0	1	?	1	3		8
SOLUBILITY		0	1	2	3	4		12
VOLATILITY		0	1	2	3	4		12
PHYSICAL STATE		0	1	2	3	4		12
ADDITIONAL POINTS FOR OTHER WASTE CHARACTERISTICS								20

NUMBER OF MISSING AND ASSUMED VALUES = _____ OUT OF 9.

PERCENTAGE OF MISSING AND ASSUMED VALUES = _____ %.

SUBTOTALS

SUBSCORE
(SITE SCORE DIVIDED BY MAXIMUM SCORE AND MULTIPLIED BY 100.)

WASTE MANAGEMENT PRACTICES								
SITE SECURITY		0	1	2	3	7		21
HAZARDOUS WASTE QUANTITY		0	1	2	3	7		21
TOTAL WASTE QUANTITY		0	1	2	3	5		15
WASTE INCOMPATIBILITY		0	1	2	3	5		15
USE OF LINERS		0	1	2	3	3		9
USE OF LEACHATE COLLECTION SYSTEMS		0	1	2	3	3		9
USE OF GAS COLLECTION SYSTEMS		0	1	2	3	2		6
USE AND CONDITION OF CONTAINERS		0	1	2	3	2		6
ADDITIONAL POINTS FOR OTHER WASTE MANAGEMENT PRACTICES								30

NUMBER OF MISSING AND ASSUMED VALUES = _____ OUT OF 8.

PERCENTAGE OF MISSING AND ASSUMED VALUES = _____ %

SUBTOTALS

SUBSCORE
(SITE SCORE DIVIDED BY MAXIMUM SCORE AND MULTIPLIED BY 100)

NUMBER OF MISSING AND ASSUMED VALUES = _____ OUT OF 31.

PERCENTAGE OF MISSING AND ASSUMED VALUES = _____ %

TOTAL SITE SCORE _____

TOTAL MAXIMUM POSSIBLE SITE SCORE _____

OVERALL SCORE _____
(TOTAL SCORE DIVIDED BY MAXIMUM SCORE AND MULTIPLIED BY 100)

data available on which to base a technical assumption. The resulting score can then be mathematically normalized so that it is comparable to scores for other sites.

In most cases, both of these methods are needed to fill data gaps. It is important that raters note which factors have been evaluated on the worksheet for rating disposal sites (Table 7.7).

Appendix for
Chapter 7
Guidance for Assessing the
Rating Factors

RECEPTORS

Population within 1000 feet is a rough indicator of the potential hazard exposure of the residential population near the site. It is measured from the site boundaries on all sides. Where only houses can be counted (e.g., from an aerial photograph), it can be computed using 3.8 individuals per dwelling unit.

Distance to nearest drinking water well is the distance between a site and the nearest downgradient building that is unlikely to be served by a public water supply (e.g., farmhouses).

Distance to the nearest off-site building is a direct indicator of potential property damage and an indirect measure of potential human exposure. The building's use is considered under the additional point system, and not in assessing this factor.

Land use/zoning is intended to indicate the nature and level of human activity in the vicinity of the site. Impending change in the zoning or land use should be noted in rating the site.

Critical environment is any environment which contains important biological resources, or which is a fragile natural setting that will suffer an especially severe impact from pollution.

PATHWAYS

Evidence of contamination indicates how confident the rater is of his or her contamination information.

This guide is for use with Tables 7.6 and 7.7.

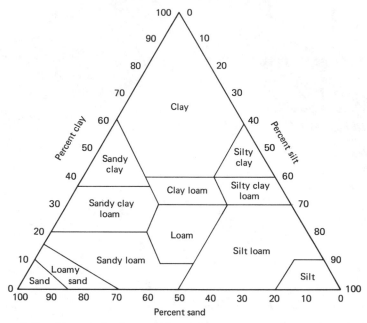

Figure 7.13. Diagram of soil composition, an important aspect of waste site evaluations.

Level of contamination signifies how readily apparent and severe a site's hazards are. In addition, it provides a rough indication of the concentration of contaminants at the site.

Type of contamination indicates what media have been contaminated by a site. Since this methodology is for land-based disposal sites, it is assumed that when air or water is contaminated, then so is the biota. And when the biota is contaminated, then so is the soil.

Distance to the nearest surface water is the shortest distance from the perimeter of the site to the nearest body of surface water (e.g., lake or stream) which periodically contains water. This measure provides an indication of the ease with which pollutants may flow overland to surface-water bodies.

Depth to ground water is measured vertically from the lowest point of the filled wastes to the highest point of the seasonal water table. This factor provides an indication of the ease with which a pollutant can contaminate ground water.

Net precipitation provides an indication of the potential for leachate generation at a site. It is equal to the mean annual precipitation minus the annual evapotranspiration.

Soil permeability indicates the speed at which a contaminant could migrate from a site. In addition, when considered with soil thickness and permeability, it can give an indication of how likely the contaminant is to be attenuated through soil filtration. The following chart gives approximate permeabilities of various soil textural classes and the level assigned to each (see Fig. 7.13).

JRB Associates Level	Textural Classes	Approximate Permeability (cm/sec)
3	Sand, loamy sand, and sandy loam	$>10^{-2}$
2	Loam, silty loam, and silt	10^{-2} to 10^{-4}
1	Sandy clay, clay loam, silty clay loam, and sandy clay loam	10^{-4} to 10^{-5}
0	Clay and silty clay	$<10^{-6}$

Bedrock permeability is another indicator of the ease and rapidity with which pollutants can migrate from a site. The following will aid the rater in evaluating this factor:

JRB Associates Level	Rock Types	Approximate Permeability (cm/sec)
3	Any well-fractured rock, carbonates and evaporites (in humid climates), and volcanic igneous rocks (except basalt)	$>10^{-2}$
2	Any slightly fractured rock, un-cemented and weakly cemented sandstones, and conglomerates	10^{-2} to 10^{-4}
1	Siltstones, moderately and well cemented sandstones, and conglomerates	10^{-4} to 10^{-5}
0	Unfractured shales, metamorphic and igneous rocks (except volcanics), and carbonates and evaporites in an arid climate	$<10^{-6}$

Depth to bedrock is the thickness of the overburden (soil and weathered rock) at the site. This provides a measure of the potential for pollutant attenuation through filtration and soil reactions when coupled with soil permeability.

WASTE CHARACTERISTICS

In order to simplify assessing a site's waste characteristics, Table 7.A.1 has been prepared listing ratings for several common chemical compounds. *Toxicity* levels are excerpted from N. I. Sax, *Hazardous Properties of Industrial Materials* and the NFPA's *Guide on Hazardous Materials*. As with all scientific data, additional verification is recommended using additional sources previously noted, or waste analysis if necessary (see Fig. 7.9).

TABLE 7.A.1.   Waste Characteristics Ratings for Several Common Chemical Compounds

Chemical/Compound	Toxicity	Persistence	Ignitability	Reactivity	Solubility	Volatility
Acetaldehyde	2	0	3	2	3	3
Acetic acid	2	0	2	1	3	1
Acetone	1	0	3	0	0	3
Aldrin	3	3	1	0	0	0
Ammonia	3	0	1	0	1	3
Aniline	3	1	2	0	2	1
Benzene	2	1	3	0	1	3
Carbon tetrachloride	2	3	0	0	0	3
Chlordane	3	3	0	0	0	0
Chlorobenzene	2	2	3	0	0	1
Chloroform	2	3	0	0	3	3
Cresol-0	3	1	2	0	2	1
Cresol-M&P	3	1	1	0	2	1
Cyclohexane	1	2	3	0	0	3
DDT	3	3	0	0	0	0
Dioxin	3	3	0	0	0	0
Endrin	3	3	1	0	0	0
Ethyl Benzene	2	1	3	0	0	1
Formaldehyde	2	0	2	0	2	3
Formic acid	3	0	2	0	3	2
Hexachlorobenzene	1	3	0	0	0	1
Hydrochloric acid	3	0	0	0	2	3
Isopropyl ether	2	1	3	1	3	3
Lindane	2	3	1	0	0	0
Methane	1	1	3	1	2	2
Methyl ethyl ketone	2	0	2	3	3	2
Methyl parathion	3	3	3	2	1	2
Naphthalene	2	1	2	0	0	1
Nitric acid	2	0	0	1	0	2
Parathion	3	3	1	0	0	0
PCB	3	3	0	0	0	0
Petroleum	1	1	3	0	0	1
Phenol	3	1	2	0	2	1
Sulfuric acid	3	0	0	2	1	0
Toluene	2	1	3	0	0	2
α-Trichloroethane	2	2	3	2	1	3
Trichlorobenzene	2	3	0	1	0	1
Xylene	2	1	3	0	0	1

Sax's Toxicity Ratings

In Section 9 in Sax (N. I. Sax, *Hazardous Properties of Industrial Materials*, 5th ed., Van Nostrand Reinhold, New York, 1979, Section 9, pp. 271–272), the following system of toxicity ratings is used to indicate the relative hazard.

U = Unknown

This rating has been assigned to chemicals for which insufficient toxicity data were available to enable a valid assessment of the hazard. These compounds usually are in one of the following categories:

1. No toxicity information found in the literature and none known to the author.
2. Limited information available based on animal experiments but in the opinion of the author this information could not be applied to human exposures. In some cases this information is mentioned so that the reader may know that some experimental work has been done.
3. Published toxicity data felt by the author to be of questionable validity.

0 = No Toxicity

This designation is given to materials which fall into one of the following categories:

1. Materials which cause no harm under any conditions of normal use.
2. Materials which produce toxic effects on humans only under the most unusual conditions or by overwhelming dosage.

1 = Slight Toxicity

1. *Acute Local.* Materials which on single exposure lasting seconds, minutes, or hours cause only slight effects on the skin or mucous membranes regardless of the extent of the exposure.
2. *Acute Systemic.* Materials which can be absorbed into the body by inhalation, ingestion, or through the skin and which produce only slight effects following single exposures lasting seconds, minutes, or hours, or following ingestion of a single dose, regardless of the quantity absorbed or the extent of exposure.
3. *Chronic Local.* Materials which on continuous or repeated exposures extending over periods of days, months, or years cause only slight and usually reversible harm to the skin or mucous membranes. The extent of exposure may be great or small.
4. *Chronic Systemic.* Materials which can be absorbed into the body by inhalation, ingestion, or through the skin and which produce only slight and usually reversible exposures extending over days, months, or years. The extent of the exposure may be great or small.

In general, those substances classified as having "slight toxicity" produce changes in the human body which are readily reversible and which will disappear following termination of exposure, either with or without medical treatment.

2 = Moderate Toxicity

1. *Acute Local.* Materials which on single exposure lasting seconds, minutes, or hours cause moderate effects on the skin or mucous membranes. These effects may be the result of intense exposure for a matter of seconds or moderate exposure for a matter of hours.

2. *Acute Systemic.*   Materials which can be absorbed into the body by inhalation, ingestion, or through the skin and which produce moderate effects following single exposures lasting seconds, minutes, or hours, or following ingestion of a single dose.

3. *Chronic Local.*   Materials which on continuous or repeated exposures extending over periods of days, months, or years cause moderate harm to the skin or mucous membranes.

4. *Chronic Systemic.*   Materials which can be absorbed into the body by inhalation, ingestion, or through the skin and which produce moderate effects following continuous or repeated exposures extending over periods of days, months, or years.

Those substances classified as having "moderate toxicity" may produce irreversible as well as reversible changes in the human body. These changes are not of such severity as to threaten life or produce serious physical impairment.

3 = SEVERE TOXICITY

1. *Acute Local.*   Materials which on single exposure lasting seconds or minutes cause injury to the skin or mucous membranes of sufficient severity to threaten life or to cause permanent physical impairment or disfigurement.

2. *Acute Systemic.*   Materials which can be absorbed into the body by inhalation, ingestion, or through the skin and which can cause injury of sufficient severity to threaten life following a single exposure lasting seconds, minutes, or hours, or following ingestion of a single dose.

3. *Chronic Local.*   Materials which on continuous or repeated exposures extending over periods of days, months, or years can cause injury to the skin or mucous membranes of sufficient severity to threaten life or cause permanent impairment, disfigurement, or irreversible change.

4. *Chronic Systemic.*   Materials which can be absorbed into the body by inhalation, ingestion, or through the skin and which can cause death or serious physical impairment following continuous or repeated exposures to small amounts extending over periods of days, months, or years.

NFPA's Toxicity Ratings

Following are the five toxicity ratings of the NFPA:

0   Materials which on exposure under fire conditions would offer no health hazard beyond that of ordinary combustible material.

1   Materials only slightly hazardous to health. It may be desirable to wear a SCBA.

2   Materials hazardous to health, but areas may be entered freely with a SCBA.

3   Materials extremely hazardous to health, but areas may be entered with extreme care. Full protective clothing including a SCBA, rubber gloves, and

boots, and bands around the legs, arms, and waist should be provided. No skin surface should be exposed.

4   A few whiffs of the gas or vapor could cause death or the gas, vapor, or liquid could be fatal on penetrating the fire fighter's normal full protective clothing which is designed for resistance to heat. For most chemicals having a Health 4 rating, the normal full protective clothing available to the average fire department will not provide adequate protection against skin contact with these materials. Only special protective clothing designed to protect against the specific hazard should be worn.

Radioactivity is evaluated in terms of background radioactivity. Background radioactivity is the level of radiation due to natural sources such as cosmic rays, building materials, and naturally radioactive materials in soil and rock. To determine background radiation levels, the investigator should use radiation detection devices, such as scintillation detectors, and measure the radiation levels at some location close to the site, but not affected by wastes deposited in the site. The elevation of the location where background radiation is measured should be similar to that of the site. This background radiation level is then compared with that found on the site and rated accordingly. (See "Radiation Hazards: Their Prevention and Control," in *Safety and Accident Prevention in Chemical Operations,* 2nd ed., H. H. Fawcett and W. S. Wood (Eds.), Wiley–Interscience, New York, 1982, Chapter 37, pp. 837–864.)

Persistence is evaluated on the biodegradability of the wastes. In rating this factor, the most persistent compound should be chosen. A guide for evaluating this factor for organics can be found in Table 7.A.2.

Ignitability provides an indication of the threat posed by fire at a site and is based on the flammability classification of the NFPA. The following chart gives the NFPA flammability characteristics and the level assigned to each:

JRB Associates Level	NFPA Level	
3	4	Very flammable gases, very volatile flammable liquids, and materials that in the form of dusts or mists readily form explosive mixtures when dispersed in air. Shut off flow of gas or liquid and keep cooling water streams on exposed tanks or containers. Use water spray carefully in the vicinity of dusts so as not to create dust clouds.
	3	Liquids which can be ignited under almost all normal temperature conditions. Water may be ineffective on these liquids because of their low flash points. Solids which form coarse dusts; solids in shredded or fibrous form that create flash fires; solids that burn rapidly, usually because they contain their own oxygen; and any material that ignites spontaneously at normal temperatures in air are in this level.

TABLE 7.A.2.   Persistence (Biodegradability) of Some Organic Compounds

Level 3: Highly Persistent Compounds

Aldrin
Benzopyrene
Benzothiazole
Benzothiophene
Benzyl butyl phthalate
Bromochlorobenzene
Bromodichloromethane
Bromoform
Bromoform butanol
Bromophenyl phenyl ether
Carbon tetrachloride
Chlordane
Chloroform
Chlorohydroxy benzophenone
Chloromochloromethane
bis-Chloroisopropyl ether
m-Chloronitrobenzene
DDE
DDT
Dibromobenzene
Dibromodichloroethane
Dibutyl phthalate
1,4-Dichlorobenzene
Dichlorodifluoroethane
Dieldrin
Diethyl phthalate
Di(2-ethylhexyl)phthalate

Dihexyl phthalate
Di-isobutyl phthalate
Dimethyl phthalate
4,6-Dinitro-2-aminophenol
Dipropyl phthalate
Endrin
Heptachlor
Heptachlor epoxide
1,2,3,4,5,7,7-Heptachloronorbornene
Hexachlorobenzene
Hexachloro-1, 3-butadiene
Hexachlorocyclohexane
Hexachloroethane
Methyl benzothiazole
Pentachlorobiphenyl
Pentachlorophenol
1,1,3,3-Tetrachloroacetone
Tetrachlorobiphenyl
Tetrachloroethane
Thiomethylbenzothiazole
Trichlorobenzene
Trichlorobiphenyl
1,1,2-Trichloroethane
Trichlorofluoromethane
2,4,6-Trichlorophenol
Triphenyl phosphate

Level 2: Persistent Compounds

Acenaphthylene
Atrazine
(Diethyl) atrazine
Barbital
Borneol
Bromobenzene
Camphor
Chlorobenzene
1,2-bis-Chloroethoxy ethane
b-Chloroethyl methyl ether
Chloromethyl ether
Chloromethyl ethyl ether
3-Chloropyridine
Di-t-butyl-p-benzoquinone

Dichloroethyl ether
Dihydrocarvone
Dimethyl sulfoxide
2,6-Dinitrotoluene
cis-2-Ethyl-4-methyl-1,3-dioxolane
trans-2-Ethyl-4-methyl-1,3-dioxolane
Guaiacol
2-Hydroxyadiponitrile
Indene
Isoborneol
Isophorone
Isopropenyl-4-isopropyl benzene
2-Methoxy biphenyl
Methyl biphenyl

TABLE 7.A.2. *(Continued)*

Methyl chloride	Nitrobenzene
Methylene chloride	Tetrachloroethylene
Methylindene	1,1,2-Trichloroethylene
Nitroanisole	Trimethyl-trioxo-hexahydro-triazine isomer

Level 1: Somewhat Persistent Compounds

Acetylene dichloride	Methane
Benzene	Methyl ester of behenic acid
Benzene sulfonic acid	Methyl ester of lignoceric acid
Butyl benzene	2-Methyl-5-ethyl-pyridine
Butyl bromide	Methyl naphthalene
e-Caprolactam	Methyl palmitate
Carbon disulfide	Methyl phenyl carbinol
o-Cresol	Methyl stearate
Decane	Naphthalene
1,2-Dichloroethane	Nonane
1,2-Dimethoxy benzene	Octane
1,3-Dimethyl naphthalene	Octyl chloride
1,4-Dimethyl phenol	Pentane
Dioctyl adipate	Phenyl benzoate
n-Dodecane	Phthalic anhydride
Ethyl benzene	Propylbenzene
2-Ethyl-n-hexane	1-Terpineol
o-Ethyltoluene	Toluene
Isodecane	Vinyl benzene
Isopropyl benzene	Xylene
Limonene	

Level 0: Nonpersistent Compounds

Acetaldehyde	Methyl benzoate
Acetic acid	3-Methyl butanol
Acetone	Methyl ethyl ketone
Acetophenone	2-Methylpropanol
Benzoic acid	Octadecane
Di-isobutyl carbinol	Pentadecane
Docosane	Pentanol
Eicosane	Propanol
Ethanol	Propylamine
Ethylamine	Tetradecane
Hexadecane	n-Tridecane
Methanol	n-Undecane

Source. E. F. Abrams, "Identification of Organic Compounds in Effluents from Industrial Sources," EPA-560/3-75-002, April 1975.

2	2	Liquids which must be moderately heated before ignition will occur and solids that readily give off flammable vapors. Water spray may be used to extinguish the fire because the material can be cooled to below its flash point.
1	1	Materials that must be preheated before ignition can occur. Water may cause frothing of liquids with this flammability rating number if it gets below the surface of the liquid and turns to steam. However, water spray gently applied to the surface will cause a frothing which will extinguish the fire. Most combustible solids have a flammability rating of 1.
0	0	Materials that will not burn.

Reactivity is a measure of the explosion threat of a site and is also based on the classification of the NFPA. See also the report by H. K. Hatayama, "A Guide for Determining Hazardous Wastes Compatibility," Grant No. R804692, U.S. Environmental Protection Agency, Municipal Environmental Research Laboratory, Cincinnati, Ohio, 1980. The following chart gives the NFPA reactivity characteristics and the level assigned to each:

JRB Associates Level	NFPA Level	
3	4	Materials which in themselves are readily capable of detonation or of explosive decomposition or explosive reaction at normal temperatures and pressures. Includes materials which are sensitive to mechanical or localized thermal shock. If a chemical with this hazard rating is in an advanced or massive fire, the area should be evacuated.
	3	Materials which in themselves are capable of detonation or of explosive decomposition or of explosive reaction but which require a strong initiating source or which must be heated under confinement before initiation. Includes materials which are sensitive to thermal or mechanical shock at elevated temperatures and pressures or which react explosively with water without requiring heat or confinement. Fire fighting should be done from an explosion-resistant location.
2	2	Materials which in themselves are normally unstable and readily undergo violent chemical change but do not detonate. Includes materials which can undergo chemical change with rapid release of energy at normal temperatures and pressures or which can undergo violent chemical change at elevated temperatures and pressures. Also includes those materials which may react violently with water or which

		may form potentially explosive mixtures with water. In advanced or massive fires, fire fighting should be done from a protected location.
1	1	Materials which in themselves are normally stable but which may become unstable at elevated temperatures and pressures or which may react with water with some release of energy but not violently. Caution must be used in approaching the fire and applying water.
0	0	Materials which are normally stable even under fire exposure conditions and which are not reactive with water. Normal fire-fighting procedures may be used.

Corrosiveness is based on the pH of the wastes and indicates the potential for damage to equipment, fixtures, and any organisms that may contact these wastes.

Solubility signifies how easily contaminants can mix with water, and thus, how readily they can migrate from the site by way of surface and ground waters. It is based on the CRC *Handbook of Chemistry and Physics* as follows:

JRB Associates Level	CRC Level
3	*V*—Very soluble, infinitely soluble, or miscible
2	*S*—Soluble
1	δ/S^h—Slightly soluble
0	*I*—Insoluble

Volatility is a measure of a disposed-material's tendency to change from a liquid, solid, or semisolid state, directly to a gaseous state under normal ambient conditions of temperature and pressure. Volatility thus provides a means of rating the potential for air-pollution problems due to the disposed material. The material's vapor pressure is the most readily available relative measure of this tendency. Unknown chemical odors at a site should be rated at least a "1" in the absence of more definitive data.

Physical state refers to the state of the wastes at the time of disposal. Gases generated by the wastes in a disposal area should not be considered in rating this factor.

WASTE-MANAGEMENT PRACTICES

Site security is an indication of what positive actions have been taken to limit the exposure of people and animals to waste-related hazards.

Hazardous-waste quantity indicates a greater potential hazard for sites with large amounts of hazardous wastes. As an aid in estimating this value, amounts

TABLE 7.A.3. Hazardous Waste Compatibility Chart

Reactivity Group No.	Reactivity Group Name												
1	Acids, Mineral, Non-oxidizing	1											
2	Acids, Mineral, Oxidizing		2										
3	Acids, Organic	G H		3									
4	Alcohols and Glycols	H	H F	H P	4								
5	Aldehydes	H P	H F	H P		5							
6	Amides	H	H GT				6						
7	Amines, Aliphatic and Aromatic	H	H GT	H		H		7					
8	Azo Compounds, Diazo Compounds, and Hydrazines	H G	H GT	H G	H G	H			8				
9	Carbamates	H G	H GT							G H	9		
10	Caustics	H	H	H		H				H G	10		
11	Cyanides	GT GF	GT GF	GT GF						G		11	
12	Dithiocarbamates	H GF F	H GF F	H GF GT		GF GT		U		H G			12
13	Esters	H	H F							H G		H	13
14	Ethers	H	H F										
15	Fluorides, Inorganic	GT	GT	GT									
16	Hydrocarbons, Aromatic	H F											
17	Halogenated Organics	H GT	H F GT							H GT	H G	H GF	H
18	Isocyanates	H G	H F GT	H G	H P					H P	H G	H P G	H G U
19	Ketones	H	H F							H G		H	H
20	Mercaptans and Other Organic Sulfides	GT GF	H F GT							H G			
21	Metals, Alkali and Alkaline Earth, Elemental	GF H F	GF H F	GF H F	GF H F	GF H F	GF H	GF H	GF H	GF H	GF H	GF GT H	GF H
22	Metals, Other Elemental & Alloys as Powders, Vapors, or Sponges	GF H F	GF H F	GF						H F GT	U	GF H	
23	Metals, Other Elemental & Alloys as Sheets, Rods, Drops, Moldings, etc.	GF H F	GF H F							H F G			
24	Metals and Metal Compounds, Toxic	S	S	S		S	S			S			
25	Nitrides	GF H F	H F E	H GF	GF H E	GF H				U	H G	U	GF H	GF H	GF H
26	Nitrites	H GT GF	H F GT	H						U			
27	Nitro Compounds, Organic	H F GT			H					H E			
28	Hydrocarbons, Aliphatic, Unsaturated	H	H F		H								
29	Hydrocarbons, Aliphatic, Saturated	H F											
30	Peroxides and Hydroperoxides, Organic	H G	H E	H F	H G	H GT	H F E	H F GT	H E	H F GT			
31	Phenols and Cresols	H	H F							H G			
32	Organophosphates, Phosphothioates, Phosphodithioates	H GT	H GT							U	H E		
33	Sulfides, Inorganic	GT GF	H HF GT	GT		H				E			
34	Epoxides	H P	H P	H P	H P	U				H P	H P	H P	H P	U
101	Combustible and Flammable Materials, Miscellaneous	H G	H F GT										
102	Explosives	H E	H E	H E						H E		H E	H E
103	Polymerizable Compounds	P H	P H	P H						P H	P H	P H	U
104	Oxidizing Agents, Strong	H GT	H GT	H F	H F	H F GT	H GT	H E	H F GT		H E GT	H E GT	H F
105	Reducing Agents, Strong	H GF	H F GT	H GF	H GF	GF H F	GF H F	H GF	H G		H GT	H F
106	Water and Mixtures Containing Water	H	H							G			
107	Water Reactive Substances	EXTREMELY REACTIVE! DO NOT MIX WITH ANY CHEMICAL OR WASTE MATERIAL! EXTREMELY REACTIVE!											

| | 1 | 2 | 3 | 4 | 5 | 6 | 7 | 8 | 9 | 10 | 11 | 12 | 13 |

Source. A Method for Determining Hazardous Wastes Compatibility, by H. K. Hatayama et al., Grant No. R804692, for Municipal Environmental Research Laboratory, Office of Research and Development, U.S.E.P.A., Cincinnati, OH 45268, 1980.

196

Reactivity Code	Consequences
H	Heat generation
F	Fire
G	Innocuous and non-flammable gas generation
GT	Toxic gas generation
GF	Flammable gas generation
E	Explosion
P	Violent polymerization
S	Solubilization of toxic substances
U	May be hazardous but unknown

Example:

H
F   Heat generation, fire, and toxic gas generation
GT

are expressed in three units below. On occasion, raters may have to convert several pieces of quantity data to a common unit in order to add them together:

JRB Associates Level	Volume (yd³)	Weight (short tons)	Drums (55 gal)
3	>2370	>2000	>650
2	1191–2370	1001–2000	326–650
1	300–1190	250–1000	80–325
0	<300	<250	<80

Total waste quantity is used as an indication of a site's size, and in conjunction with *hazardous-waste quantity* as an estimate of what proportion of the total wastes are hazardous. Additional volume units are given below. Raters can estimate a

Figure 7.14. Pollulert Fluid Detection System Continuous Electronic Monitoring System for the detection of hazardous liquid leaks and spills. For additional information, contact Pollulert Systems, P.O. Box 706, Indianapolis, Indiana 46206.

site's total waste quantity by multiplying the area of the site (in acres) by the average thickness of the wastes (in feet) to obtain acre-feet:

JRB Associates Level	Acre-Feet	Cubic Yards
3	>250	>400,000
2	100–250	161,000–400,000
1	10–100	16,000–161,000
0	<10	<16,000

Waste incompatibility can lead to explosion, fire, and the production of toxic fumes. A table of incompatible wastes is included in Table 7.A.3 as an aid in evaluating this factor.

Use of liners provides an indication of improvement on the impermeability of natural soils occurring at the site. Although there are wide variations in liner quality and durability, the presence of a liner indicates that an effort was made to reduce or eliminate migration of contaminants, which is a positive factor in the evaluation of disposal sites. The scale shown in Table 7.A.1 assumes that the site being rated contains organics, which represents a "worst case" scenario. Sites without organics can be rated accordingly.

Use of leachate collection systems indicates that action has been taken to prevent ground-water contamination problems. Hence, it should be viewed as a positive factor in the rating of disposal sites. Raters must judge on a case-by-case basis what constitutes "adequate" collection and treatment (see Fig. 7.4 and Fig. 7.14).

Use of a gas collection system is another positive factor that indicates an effort is being made to control the release of pollutants from a site. Sampling and rater judgment may be necessary to determine the adequacy of collection and treatment.

Use and condition of containers provides an indication of whether wastes were dumped in bulk or in containers, and the integrity of the containers used. Containers retard the release of pollutants to the environment provided containers are intact.

ALTERNATIVE DISPOSAL MODES

While the immediate out-of-pocket relatively low cost of landfilling will doubtlessly encourage landfill operations for the immediate future, the limitations of landfills and alternative handling and disposal methods should be recognized. We have noted the possibility of recycling and other approaches, including incineration. Bruce Piasecki, in an article in *Science 83,* **4,** No. 7, 76–81 (September 1983), has called attention to the serious physical limitations of even secure landfills. He notes also the loopholes in the present laws, such as the "small generator" exemption from recordkeeping or monitoring for the 700,000 small firms responsible for 92% of all U.S. toxic-waste production. Another loophole, the boiler exception, permits waste solvents, creosote, and other hazardous materials to be burned as fuel. One estimate states that 1000 boilers in the United States may be fueled by hazardous waste, and that only 20 out of 100 toxic-fuel-burning boilers checked were being monitored for efficiency in destruction of waste. Piasecki notes that there are technologies either available now or in the near future which will serve as alternatives to landfills. The following lists some technologies:

Bacteria which can digest oil has been patented and others which can destroy dioxins and other toxins.

Water heated to a "supercritical" point by intense heat and pressure can dissolve many toxic wastes.

Molten salt incineration is possible and practical by treatment with sodium carbonate at 1650°F (900°C) (considerably lower than that needed for complete incineration of most waste materials).

PCB trucks have been developed by SunOhio Company, as well as by the Acurex Corporation, making possible "on-scene" treatment of PCBs in utility transformers and electrical capacitors.

The plasma arc, which uses electricity between two electrodes to heat air inside a chamber to 45,000°F (25,000°C), transforms the air into a plasma. Toxic waste is fed into the center of the chamber where high-energy electrons bombard the material, breaking molecular bonds and reducing the material to basic elements. The newest version can treat a continuous stream of wastes.

In order to ensure that existing "hazardous-waste-secure" landfills are, in fact, retaining their ability to contain, monitoring wells to observe any leachate is very important over time. A dedicated system of monitoring such wells has been developed by Q.E.D. of Ann Arbor, Michigan, and was described in detail in the paper "Selection of Cost-Effective Compliance Monitoring Equipment," by W. D. Dickinson and D. Mioduszewski (both of Q.E.D.) which was published in the Proceedings of the First Hazardous Materials Management Conference held in Philadelphia, Pennsylvania on July 12, 1983 to July 14, 1983. Proceedings are available from Tower Conference Publishing Company, 143 North Hale Street, Wheaton, IL 60187. See also Fig. 7.14 for another system for monitoring liquids in leachate systems.

HOUSEHOLD WASTES

One of the loopholes in the RCRA law is that it does not include hazardous wastes from consumer products in the home. Although the individual items may seem small compared to industrial waste, the effect of some household items may be significant, especially where the wastes reach either a landfill, a stream of water, or an incinerator. The Audubon Naturalist Society is concerned that the role of household wastes is not fully recognized. By using a similar project conducted in Seattle, Washington by the Metropolitan Seattle Metro, the Audubon Naturalists have attempted, by education and affirmative action, to create such an awareness. Even though hazardous materials in the home are unregulated, once they are collected, stored, transported, treated, or disposed of they are subject to local, state, and federal laws. These laws, as noted, require a manifest with information on the chemical composition of the waste.

The Audubon Naturalist Society states the problem as follows:

> Many household, auto, and garden products are harmful if used or disposed of improperly. Many contain chemicals which are toxic and/or caustic or acid, causing burns to people, plants, and animals. Under the Federal Clean Water Act of 1977, more than 100 of these toxic substances, or toxicants, commonly used by industry, commercial businesses, and households, have been listed as "priority pollutants." Toxicants may be poisonous even in small amounts, and can often accumulate in the environment. It is the continuous release of small amounts of toxic substances that can, with time, create major problems.
>
> 1. The cost of sewage treatment can increase dramatically if toxicants are in high concentrations in the waste stream.
>
> 2. Many toxicants enter our waters from sources outside the sewerage system.
>
> 3. The cost of cleaning up waters, once polluted with toxicants, can be astronomical. Therefore, control must begin at the source.

General Tips on Household Chemicals

USE

Be aware of the uses and dangers of products. If directions for use and disposal are unclear, contact the manufacturer or dealer before using or disposing.

Keep unused products in their original containers for directions to follow, and a list of contents for reference in case of accidental poisonings. This, of course, is especially important where children or pets may have access to the materials.

Never mix different products: explosive or poisonous chemical reactions may occur. The classic case of acidic toilet-bowl cleaner being mixed with bleach producing chlorine and other hazardous gases is well known but infrequently referred to.

Buy the smallest amount of substance that you need, if possible, to reduce storage problems and disposal questions.

Read and observe all label instructions; the above precautions are often overlooked.

DISPOSAL

Products containing harmful chemicals should not be poured or flushed down the drain unless you are fully aware of the potential effect on the sewer system. Many substances disturb or otherwise disrupt septic tanks or pass through systems directly into the outfall of rivers.

Excessive amounts of products containing hazardous materials should be disposed of by professionals who are certified and licensed to do so by the local or state authorities. Sharing surplus with others who can profitably use it (such as fertilizer, oil, or thinners) is another approach.

PESTICIDES AND HERBICIDES

Pesticides and herbicides are designed to produce specific effects on plants and animals, and are poisonous, to some degree, to people, animals, and other wildlife. Among pesticides now banned or restricted which merit special attention are the following:

DDT
Aldrin
Dieldrin
Chlordane
DBCP
Heptachlor
Lindane
Kepone
Mirex
Silvex
2,4,5-T
Toxaphene

After using a pesticide or herbicide, in a garden or farm area, take care to avoid excessive overwatering, since the pesticide or herbicide may run off with excess water into nearby streams, and hence cause pollution.

PAINT PRODUCTS AND PRESERVATIVES

Paints, lacquers, paint thinners and strippers, brush cleaners, wood preservatives, and turpentine are examples of these products. They may contain a variety of toxic substances, such as benzene, toluene, lead (in older paint), methyl chloride, pentachlorophenol, and trichloroethane, as well as trichloroethylene.

Solvents, paint thinners, and wood preservatives, as well as paints, should not be dumped into a sewer system. Incineration in a small, carefully controlled open fire in an open area away from personnel is one alternative; recycling by possible reuse is another.

CLEANERS AND DETERGENTS

Deodorizers, car cleaners, polishes, spot removers, prewashes, and drain cleaners are examples of these substances. They may contain lye (sodium and/or potassium hydroxide), sodium hypochlorite, petroleum distillates, naphthas, trichloroethylene and trichloroethane, phenols, dichlorobenzene, and naphthalene (the last two are commonly referred to as "moth balls").

Many are acute or chronic poisons, accumulating to toxic levels in people, fish, and wildlife.

As alternatives, the use of less toxic or nontoxic substances or operations should be considered. For example, in place of a caustic drain cleaner, a long "snake" (a metal rod which is flexible), together with a plunger and hot water, is often effective. Commercial services which use mechanical devices are available for opening the drain. For general cleaning, bicarbonate of soda is a relatively bland, yet effective, substitute. Biodegradable and low-phosphate products should be used where possible. A stiff brush, or a wire brush, is often effective with soap.

AUTOMOTIVE WASTES

Waste crankcase drainings constitute the largest volume of wastes from automobiles, and contain dissolved petroleum hydrocarbons, lead, zinc, and other contaminants. Waste oil is toxic to humans, fish, and animals. Waste oil should not be dumped into sewers, drains, or streams, but placed in a container with tightfitting lid and taken to a service station. Service stations have arrangements with transporters and recycling firms to remove and recycle the oil after filtration and, if needed, redistillation.

Motor coolant, such as ethylene glycol, is toxic to humans, fish, and animals. It should not be dumped or washed into drains. As with oils, service stations will receive such waste and incorporate it into their system.

Additional information on disposal of domestic wastes may be obtained from the Audubon Naturalist Society Conservation Department at (301) 652-9188; from the U.S. Environmental Protection Agency Toxic Substances Hotline at (800) 424-9346; and from the Poison Control Center Hotline (800) 942-2414, or your local hospital poison-control center (usually listed in the front of the telephone directory).

One recent publication notes that a growing number of individuals have suffered an immune-system breakdown through sensitivity to everyday chemicals and foods. For allergic persons, the book *How To Live With the New 20th Century Illness— A Resource Guide for Living Chemically-Free,* by Linda and Milton Weiss, P.O. Box 64, Franklin, Michigan 48025 (reviewed in *the Washington Times Magazine,* September 1, 1983, pp. 3D–14D), may be of interest and possible value.

A recent approach to the disposal of known or potential carcinogens consists of inactivation methods. In the July 1983 issue of *Lab Safety Views,* published by the Division of Safety of the National Institutes of Health, Eric B. Sansone notes that

while biological agents may be rapidly and completely destroyed using formaldehyde, ethylene oxide, or by autoclaving, few procedures exist for the destruction
of highly toxic chemicals. Most destruction methods that have been reported to be
effective have not been validated. Such methods must be used with caution, if at
all.

Sansone reports that, in light of this problem, for the last several years, the
Division of Safety has sponsored the Environmental Control and Research Program
at the NCI-Frederick Cancer Research Facility and the International Agency for
Research on Cancer to seek in-lab (i.e., at the bench) methods for the decontamination and destruction of chemical carcinogens in laboratory wastes and to validate
those methods by collaborative studies.

These investigations have so far resulted in the successful development, validation, and publication of decontamination and destruction techniques for aflatoxins,
N-nitrosamines, and polycyclic aromatic hydrocarbons. Methods that may be applied to hydrazines, nitrosamides, aromatic amines, and certain antineoplastic drugs
are the subjects of current research.

Sansone invites all interested researchers to contribute to the list of suggested
methods for the destruction and disposal of laboratory waste containing chemical
carcinogens and to participate in collaborative validation studies. In developing
new laboratory destruction and disposal methods of waste, the methods must:

Completely destroy the toxic material, with no detectable material remaining in
any reaction mixture.

Be safe and irreversible.

Produce only innocuous materials (end products).

Be simple to verify in terms of effectiveness.

Require equipment and chemicals that are readily available, inexpensive, and
simple to use.

Not require elaborate operations (such as distillation or extraction) nor reagents
that have shelf-life limitations.

Be easy to reliably perform and require minimal time.

Be applicable to a wide variety of media commonly used in the research laboratory.

The references cited for the methods which have been developed to date are:

Castegnaro et al., *Laboratory Decontamination and Destruction of Aflatoxins
B_1, B_2, G_1, G_2 in Laboratory Wastes,* International Agency for Research on
Cancer, Lyon, France, 1980.

Castegnaro et al. *Decontamination and Destruction of Carcinogens in Laboratory
Wastes: Some N-Nitrosamines.* International Agency for Research on Cancer,
Lyon, France, 1982.

Castegnaro, M., *Laboratory Decontamination and Destruction of Carcinogens
in Laboratory Wastes: Some Polycyclic Aromatic Hydrocarbons,* International
Agency for Research on Cancer, Lyon, France, 1983.

Figure 7.15. Protesters came with specific law citations when demonstrating against alleged well contamination. Courtesy of Battle Creek, Michigan Enquirer and Ms. Jean Miller.

Additional information about this program is available in the United States from Dr. Eric B. Sansone, Director, Environmental Control and Research Program, NCI-Frederick Cancer Research Facility, Frederick, MD 21701, or *Lab Safety Views,* Division of Safety, National Institutes of Health, Building 13, Room 2E43, Bethesda, MD 20205.

Summary:

The intent of the Congress is that hazardous chemicals, including waste products of all kinds, should be controlled, to the end that ill effects are not produced. Sufficient interest has been generated in the RCRA and SUPERFUND programs, which we have discussed, to make for a well-informed public. As an example, Fig. 7.15 illustrates the specific citations which demonstrators claim are being violated, causing contamination of their domestic water wells. More such "informed protests" can be expected in the future, if the hazardous waste and other hazardous materials are not adequately controlled.

REFERENCES

1.   Resource Conservation and Recovery Act, Public Law 94-580, October 21, 1976.
2.   Waste Alert: (17 articles), *EPA J.* **5,** No. 2 (February 1979).

3. "Waste Disposal Site Survey," a report together with additional and separate views by the Subcommittee on Oversight and Investigations of the Committee on Interstate and Foreign Commerce, House of Representatives, 96th Cong. 1st sess., Committee Print 96-IFC 33, October 1979.

4. "Six Case Studies of Compensation for Toxic Substances Pollution: Alabama, California, Michigan, Missouri, New Jersey and Texas," a report prepared under the supervision of the Congressional Research Service of the Library of Congress for the Committee on Environment and Public Works, U.S. Senate, 96th Cong., 2d sess., Serial No. 96-13, June 1980.

5. "Health Effects of Toxic Pollution: A Report from the Surgeon General and A Brief Review of Selected Environmental Contamination Incidents with a Potential for Health Effects," reports prepared by the Surgeon General, Department of Health and Human Services and the Congressional Research Service, Library of Congress, for the Committee on Environment and Public Works, U.S. Senate, 96th Cong., 2d sess., Serial No. 96-15, August 1980.

6. "Interim Report on Ground Water Contamination: Environmental Protection Agency Oversight," Twenty-Fifth Report by the Committee on Government Operations, Union Calendar No. 874, House Report No. 96-1440, September 30, 1980.

7. "Management of Hazardous Chemical Wastes," Smithsonian Science Information Exchange, Washington, D.C., August 1979.

8a. "Test Methods for Evaluating Solid Waste (SW-846), Waste Analysis Program," Office of Solid Waste, U.S. Environmental Protection Agency (WH-565), Washington, D.C., 1980.

 b. "Technical Update, Physical/Chemical Methods, Test Methods for Evaluating Solid Waste, Revision A," SW-846, Office of Water and Waste Management, U.S. Environmental Protection Agency, Washington, D.C., August 8, 1980.

9a. R. Fung, "Protective Barriers for Containment of Toxic Materials," Pollution Technology Review No. 66, Noyes Data Corp., Park Ridge, N.J., 1980.

 b. D. J. DeRenzo, "Biodegradation Techniques for Industrial Organic Wastes, Pollution Technology Review No. 65, Noyes Data Corp., Park Ridge, N.J., 1980.

 c. H. H. Fawcett, "Experimental Facility for the Disposal of Waste Solvents," Report No. 61-GP-216, General Electric Research Laboratory, Schenectady, New York, 1962.

 d. R. D. Ross, "Incineration—A Positive Solution to Hazardous Waste Disposal, presented to American Chemical Society Division of Chemical Health and Safety, Washington, D.C., September 10, 1979.

10. "Methodology for Rating the Hazard Potential of Waste Disposal Sites," prepared by JRB Associates for the Office of Research and Development, U.S. Environmental Protection Agency, Washington, D.C., May 5, 1980.

11. A. DeBruin, *Biochemical Toxicology of Environmental Agents,* Elsevier, Amsterdam, 1976.

12. H. H. Fawcett, "Safety and Industrial Hygiene in the Laboratory," *Chem. Eng. News* **30** (251), 2588–2591 (June 23, 1952).

13. NFPA, *Manual of Hazardous Chemical Reactions,*" 5th ed., 491-M, National Fire Protection Association, Boston, Mass., 1975.

14. M. D. Morrissette, "Hazard Evaluation of Chemicals for Bulk Marine Shipment," **3,** No. 1, 33–48 (1979) and Navigation and Vessel Inspection Circular No. 4-75, U.S. Coast Guard, December 1975.

15. Leslie Bretherick, *Handbook of Reactive Chemical Hazards,* 2nd ed. Butterworth, London, 1979.

16. *Chemical Boobytraps,* 16-mm sound color movie, 10 min, 1959. (Available from General Electric Co., Audiovisuals Section, Scotia, New York.)

17. H. K. Hatayama, et al., "A Method for Determining Hazardous Wastes Compatibility," Grant No. R804692, Municipal Environmental Research Laboratory, Office of Research and Development, U.S. Environmental Protection Agency, Cincinnati, Ohio, 1980.

BIBLIOGRAPHY

"Acute and Chronic Toxicity of HCN to Fish and Invertebrates," NTIS, PB-293 047/7BE, 1979.

Adams, J., and Detjen, J., "Bibliography of Chemical Waste Disposal 1969–1979, Developed during research for 'Warning: Toxic Waste,' " *Louisville Courier-Journal,* November 25, 1979.

"Aerial Reconnaissance of Hazardous Substances Spills and Spill-Threat Conditions," National Technical Information System, PB 294 980/8BE, Springfield, Virginia, 1979.

Alter, H., and Dunn, J. J., *Solid Waste Conversion to Energy: Current European and U.S. Practice,* Marcel Dekker, New York, 1980.

American Chemical Society Division of Chemical Health and Safety, 183rd National Meeting, March 28, 1982–April 2, 1982, Las Vegas, Nevada.

   a. Jameson, C. W., and Walters, D. B., co-chairman, Symposium on Chemistry and Safety for Toxicity Testing of Environmental Chemicals (36 papers). (See Meeting Abstracts for details.)

   b. Taggart, W. P., chairman, Symposium on Laboratory Waste Disposal (12 papers). (See Meeting Abstracts for details.)

   c. Young, J. A., chairman, Symposium on Fire Toxicity (5 papers). (See Meeting Abstracts for details.)

American Chemical Society Division of Chemical Health and Safety, 186th National Meeting, August 29, 1983–September 2, 1983, Washington, D.C.:

   a. Brown, J., Prof., chairman, "Tort Law" (four-paper session).

   b. Gerlovich, J. A., "State Plan for Elimination of Unwanted/Dangerous Chemicals from School Science Labs."

   c. Kizer, F. D., "Safety in the School Science Laboratory."

   d. Schiller, J. E., "A New Reaction of Cyanide with Peroxide and Thiosulfate at PH 7 to 9."

   e. Symposia, Laboratory Waste Management: Today and Tomorrow, (joint with ACS task force on RCRA) (13 papers).

   f. Symposium on Chlorinated Dioxins and Dibenzofurans in the Total Environment (Joint with Division of Analytical Chemistry).

   g. Waters, P. F., "Large Scale Disposal of Toxic Wastes."

   h. Wodka, S., "Toxic Torts."

"Aquatic Toxicology (Third Symposium)," STP 707, American Society for Testing and Materials, Philadelphia, Pennsylvania, 1980.

"Aquatic Toxicology," STP 667, American Society for Testing and Materials, Philadelphia, Pennsylvania, 1980.

"Aquatic Toxicology and Hazard Evaluation," STP 634, American Society for Testing and Materials, Philadelphia, Pennsylvania, 1980.

Azad, H. S., *Industrial Wastewater Management Handbook,* McGraw-Hill, New York, 1976.

"Bacterial Indicators/Health Hazards Associated with Water," STP 635, American Society for Testing and Materials, Philadelphia, Pennsylvania, 1980.

Bennett, G., Feates, F. S., Wilder, I., eds., *Hazardous Materials Spill Handbook*, McGraw-Hill, N.Y. 1984.

Berry, E. E., and MacDonald, L. P., "Experimental Burning of Used Automotive Crankcase Oil in a Dry-Process Cement Kiln," *J. Hazard. Mater.* **1**, No. 2, 137–156 (1976).

"Biological Data in Water Pollution Assessment: Quantitative and Statistical Analyses," STP 652, American Society for Testing and Materials, Philadelphia, Pennsylvania, 1980.

"Biological Monitoring of Water and Effluent Quality," STP 607, American Society for Testing and Materials, Philadelphia, Pennsylvania, 1980.

Brown, M., *Laying Waste,* Pantheon Press, New York, 1980.

Buikema, A. L., Jr., and Cairns, John, Jr., "Aquatic Invertebrate Bioassays," ASTM/STP 715, American Society for Testing and Materials, Philadelphia, Pennsylvania, 1980.

Callely, A. G., *Treatment of Industrial Effluents,* Halsted Press, John Wiley, New York, 1977.

Chappell, C. L., and Wellets, S. L., "Some Independent Assessments of the SEALOSAFE/STABLEX Method for Toxic Waste Treatment," *J. Hazard. Mater.* **3,** No. 4, 285–292 (1980).

"Characteristics of Hazardous Waste Transportation and Economic Impact Assessment of Hazardous Waste Transportation Regulations," National Technical Information Service, PB-296 497/1BE, Springfield, Virginia, 1979.

Chen, E., *PBB: An American Tragedy,* Prentice-Hall, Englewood Cliffs, N.J., 1979.

Cheremisinoff, P. N., and Morresi, A. C., *Energy from Solid Wastes,* Marcel Dekker, New York, 1976.

"Control of Hazardous Material Spills," Proceedings of the 1980 National Conference on Control of Hazardous Material Spills, May 13, 1980—May 15, 1980, Louisville, Kentucky, sponsored by the U.S. Environmental Protection Agency, U.S. Coast Guard, and Vanderbilt University.

Cook, R., *Fever,* Putnam, New York, 1982.

Cope, F., Karpinski, G., Pacey, J., and Steiner, L., "Use of Liners for Containment at Hazardous Waste Landfills," *Pollution Engineering,* Vol. XVI, No. 3, 22–32 (March 1984).

Daley, P. S., "Military Marches Toward New Horizons in Pollution Control," *Pollution Engineering,* Vol. XVI, No. 2, 30–33 (Feb. 1984).

Davies, D. R., and Mackay, G. A., "Recent Developments in the Transport of Liquid Wastes," *J. Hazard Mater.* **1,** No. 3, 199–214 (1976).

Dawson, G. W., "The Acute Toxicity of 47 Industrial Chemicals to Fresh and Saltwater Fishes," *J. Hazard. Mater.* **1,** No. 4, 303–318 (1977).

Dawson, G. W., "Treatment of Hazardous Materials Spills in Flowing Streams with Floating Mass Transfer Agents," *J. Hazard. Mater.* **1,** No. 1, 65–82 (1975).

"Destroying Chemical Wastes in Commercial-Scale Incinerators," SW 122, 6 vols., U.S. Environmental Protection Agency, Washington, D.C., 1977.

"Estimating the Hazard of Chemical Substances to Aquatic Life," STP 657, American Society for Testing and Materials, Philadelphia, Pennsylvania, 1980.

Environmental Law Handbook, 6th ed. Government Institutes, Inc., Washington, D.C., 1979.

"Environmental Statutes, 1981," Government Institutes, Inc., Washington, D.C., 1981.

Epstein, S. S., Brown, L. O., and Pope, C., *Hazardous Waste in America,* Sierra Club Books, San Francisco, 1982.

Finley, S., *Hazardous Waste Options,* 16-mm sound movie, 28 min., 1981. (Available from Stuart Finley, Inc., 3428 Mansfield Rd., Falls Church, VA 22041.)

Fjerdingstod, E., "Sulfur Bacteria," STP 650, American Society for Testing and Materials, Philadelphia, Pennsylvania, 1980.

Gilmore, W. R., *Radioactive Waste Disposal—Low and High Level,* Noyes Data Corp., Park Ridge, N.J., 1977.

Hanson, D. J., Progress under Superfund criticized, defended, *Chem. Engr. News* **60,** No. 23, 10–15 (June 7, 1982).

"Hazardous and Industrial Waste Testing: Fourth Symposium," ASTM Proceedings, May 2, 1984—May 4, 1984, Stouffer's National Center Hotel, Arlington, Virginia (James K. Petros, Jr., chairman, Union Carbide, R & D Department, P.O. Box 8361, South Charleston, WV 25303.)

"Hazardous Materials Spill Monitoring Safety Handbook and Chemical Hazard Guide: Part A," PB 295 853/6BE; "Part B," PB 295 854/4BE, National Technical Information Service, 1979.

"Hazardous or Difficult to Handle Waste Survey Report," PB-292-968/5BE, National Technical Information Service, 1979.

"Hazardous Substances Summary and Full Development Plan," PB-289 923/5BE, National Technical Information Service, 1979.

"Hazardous Waste Site Investigation Training Manual," prepared by the FIT National Project Management Office, Ecology and Environment, Inc., Arlington, Virginia under EPA contract No. 68-01-6056 (TDD: HQ-8008-04), 1980.

"Hazardous Waste Training Bulletin for Supervisors" (monthly), 177 Greenwich Avenue, Stamford, CT 06904 (Bureau of Business and Law, Inc.).

Jones, C. J., "Absorption of Some Toxic Substances by Waste Components," *J. Hazard. Mater.* **2,** No. 3, 219–226 (1978).

Jones, C. J., "An Investigation of the Degradation of Some Dry Cell Batteries Under Domestic Waste Landfill Conditions," *J. Hazard. Mater.* **2,** No. 3, 259–290 (1978).

Jones, C. J., "An Investigation, of the Evaporation of Some Volatile Solvents from Domestic Waste Landfill Conditions," *J. Hazard. Mater.* **2,** No. 3, 235–252 (1978).

Jones, C. J., "Evaporation of Mercury from Domestic Waste Leachate," *J. Hazard. Mater.* **2,** No. 3, 253–258 (1978).

Jones, C. J., "The Combustion and Pyrolysis of Some Halogenated Organic Compounds in a Laboratory Tube Furnace," *J. Hazard. Mater.* **2,** No. 3, 291–296 (1978).

Jones, C. J., "The Leaching of Some Halogenated Organic Compounds from Domestic Waste," *J. Hazard. Mater.* **2,** No. 3, 1978.

Jones, C. J., "The Ranking of Hazardous Materials by Means of Hazard Indices," *J. Hazard. Mater.* **2,** No. 4, 363–389 (1978).

Korb, L. J. (Letter to the Editor), "Policing up the Area," *Wall Street Journal,* August 9, 1983, p. 33.

Krenkel, P. A., and Novotny, V., *Water Quality Management,* Academic Press, New York, 1980.

Lazar, E. C., "Damage Incidents from Improper Land Disposal," *J. Hazard. Mater.* **1,** No. 2, 157–164 (1976).

"Liners for Sanitary Landfills and Chemical and Hazardous Waste Disposal Sites," PB-293 335/6BE, National Technical Information Service, 1979.

Lowrance, W. W. (Ed.), *Assessment of Health Effects at Chemical Disposal Sites,* William Kaufmann, Inc., Los Altos, Calif., 1981.

Lund, N.H., "The Army's Environmental Quality Consulting Firm," *Pollution Engineering,* Vol. XVI, No. 2, 34–37 (Feb. 1984).

"Measurement of Organic Pollutants in Water and Wastewater," STP 686, American Society for Testing and Materials, Philadelphia, Pennsylvania, 1980.

"Metal Bioaccumulation in Fishes and Aquatic Invertebrates, A Literature Review," PB-290 659/2BE. National Technical Information Service.

"Methodology for Biomass Determinations and Microbial Activities in Sediments," STP 673, American Society for Testing and Materials, Philadelphia, Pennsylvania, 1980.

"Methods and Measurements of Periphyton Communities: A Review," STP 690, American Society for Testing and Materials, Philadelphia, Pennsylvania, 1980.

Meyer, E., *Chemistry of Hazardous Materials,* Prentice-Hall, Englewood Cliffs, N.J., 1976.

Milley, J. N., "What Really Happened at EPA," *Readers Digest,* July 1983, pp. 57–64.

National Materials Advisory Board, *Management of Hazardous Industrial Wastes: Research and Development Needs,* National Research Council, Washington, D.C., 1983.

National Wildlife Federation, "The Toxic Substances Dilemma, A Plan for Citizen Action," National Wildlife Federation, Washington, D.C., 1980.

"Native Aquatic Bacteria: Enumeration, Activity and Ecology," STP 695, American Society for Testing and Materials, Philadelphia, Pennsylvania, 1980.

Neissen, W., *Combustion and Incineration Processes,* Marcel Dekker, New York, 1978.

"News Focus: Senator Stafford: Linchpin For Environmental Legislation," *Chem. Eng. News,* 8–12 (September 5, 1983).

"Oil and Hazardous Materials: Emergency Procedures in the Water Environment, CWR 10-1," U.S. Federal Water Pollution Control Administration, Northeast Region, North Atlantic Water Quality Management Center, Edison, N.J., October, 1968.

Parves, D., *Trace-Element Contamination of the Environment,* Elsevier, Amsterdam and New York, 1977.

Patrick, P. K., "Treatment and Disposal of Hazardous Wastes in Western Europe," *J. Hazard. Mater.* **1,** No. 1, 45–58 (1975).

Powers, P. W., *How to Dispose of Toxic Substances and Industrial Wastes,* Noyes Data Corp., Park Ridge, N.J., 1976.

Proceedings of 2nd Annual HazMat 84 Hazardous Materials Conference, June 5–7, 1984, Philadelphia Civic Center, Philadelphia, PA., available from Tower Conference Management Company, 331 West Wesley Street, Wheaton, Illinois 60187.

Publications available from U.S. Environmental Protection Agency Office of Water and Waste Management, Washington, D.C. 20640:

"Cleaning Up America's Dumps: State Solid Waste Management Plans Under RCRA," SW-831, 1980.
"Everybody's Problem: Hazardous Waste," SW-826, 1980.
"Hazardous Waste Facts," SW-737, 1979.
"Operating A Recycling Program: A Citizen's Guide," SW-830, 1980.
"Public Participation Requirements for Federal, State, and Local Agencies," SW-833, 1980.
"Solid Waste Management Programs under RCRA," SW-834, 1980.
"Source Separation/Waste Reduction," SW-832, 1980.
"Waste Alert! A Citizen's Introduction to Public Participation in Waste Management," SW-800, 1979.
"What Is Waste Alert?" SW-814, 1979.

Quinlivan, S. C., et al., "Sources, Characteristics and Treatment and Disposal of Industrial Wastes Containing Hexachlorobenzene," *J. Hazard. Mater.* **1,** No. 4, 343–360 (1977).

"Remedial Action at Waste Disposal Sites," EPA-625/6-82-006, June 1982 (available from U.S. Environmental Protection Agency, Cincinnati, OH.

Renolds, J. M., "Controlling the Use of Hazardous Materials in Research and Development Laboratories," *J. Hazard. Mater.* **2,** No. 4, 299–308 (1978).

"Report of the Interagency Ad Hoc Work Group for the Chemical Waste Incinerator Ship Program," U.S. Environmental Protection Agency, U.S. Department of Commerce Maritime Administration, U.S. Coast Guard, and National Bureau of Standards, September 1980.

Resource Recovery and Conservation (An International Journal) Quarterly, Elsevier, Amsterdam and New York.

Rogers, H. W., et al., "Problems of Waste Chemical Handling at a Large Biomedical Research Facility," *J. Hazard. Mater.* **4,** No. 2, 191–205 (1980).

Rohlich, G. A., "Summary Report, Drinking Water and Health," Committee on Safe Drinking Water, National Research Council, Washington, D.C., 1977.

Roulier, M. H., et al., "Current Research on the Disposal of Hazardous Wastes," *J. Hazard. Mater.* **1,** No. 2, 59–64 (1975).

Ryan, M. J., "Air Force Environmental Quality R&D," *Pollution Engineering,* Vol. XVI, No. 2, pp. 38–42, Feb. 1984.

Schnabel, G. A., Christofano, E. E., and Harrington, W. H., "Safety and Industrial Hygiene During Investigations of Uncontrolled Waste Disposal Sites," pp. 107–110, and Blackman, W. C., Jr., "Environmental and Safety Procedures for Evaluation of Hazardous Waste Disposal Sites," pp. 91–106; in *Management of Uncontrolled Hazardous Waste Sites,* U.S. EPA National Conference, October 15, 1980–October 17, 1980, Washington, D.C.

Silvestri, A., et al., "Development of a Kit for Detecting Hazardous Material Spills in Waterways,"

EPA-600/2-78-055, Environmental Protection Technology Series, U.S. Environmental Protection Agency, Industrial Environmental Research Laboratory, Cincinnati, Ohio, 1978.

Sittig, J., *Toxic Metals, Pollution Control and Waste Protection,* Noyes Data Corp., Park Ridge, N.J., 1976.

Small, W. E., *Third Pollution: The National Problem of Solid Waste Disposal,* Praeger, New York, 1970.

Smith, A. J., *Managing Hazardous Substances Accidents,* McGraw-Hill, New York, 1981.

Smith, W. E., and Smith, A. M., *Minamata, The Story of the Poisoning of a City (with Mercury) and of the People Who Choose to Carry the Burden of Courage,* Alskog-Sensorium Book, Holt, Rinehart & Winston, New York, 1975.

Spiro, T. G., and Stigliani, W. M., "The Chemical Bases of Environmental Issues," ACS Audio Course (cassettes with manual), American Chemical Society, Washington, D.C., 1980.

Taylor, J. M., and Parr, J. F., "Considerations in the Land Treatment of Hazardous Wastes: Principles and Practices," Biological Waste Management and Organic Resources Laboratory, Agricultural Environmental Quality Institute, Science and Education Administration, U.S. Department of Agriculture, Beltsville, Maryland.

Thorne, P. F., "The Dilution of Flammable Polar Solvents by Water for Safe Disposal," *J. Hazard. Mater.* **2,** No. 4, 321–331 (1978).

Throop, W. M., "Alternate Methods of Phenol Wastewater Control," *J. Hazard. Mater.* **1,** No. 4, 319–330 (1977).

Valovic, T. S., "Hazardous Waste Management: Overview of Proposed EPA Regulations and Their Impact Upon Industry," CU Risk Management, Inc., Boston, Massachusetts, April 1980.

Voss, G., "Energy and Resource Recovery from Solid Waste," AAAS Intergovernmental Research and Development Project, September 24, 1979–September 26, 1979, Lanham, Maryland, American Association for the Advancement of Science, Washington, D.C.

Wakeman, R. J., "Filtration Post-Treatment Processes," *Chemical Engineering Monographs,* Vol. 2, Elsevier, Amsterdam and New York, 1975.

Wexley, P., *Information Resources in Toxicology,* Elsevier/North Holland, Amsterdam, 1982.

Willmann, J. C., "Case History: PCB Transformer Spill," *J. Hazard. Mater.* **1,** No. 4, 361–372 (1977).

Wulfinghoff, Max, "Disposal of Process Wastes, Liquids, Solids, Gases," a symposium presented at the ACHEMA Meeting, 1964, Frankfurt/Main, Germany, Chemical Publishing Co., New York, 1968 (translation).

RCRA and SUPERFUND

Alsop, R., "Chemical, Gas Firms Seek Legal Immunity For Helping Clean Up Hazardous Spills," *Wall Street Journal,* 29 (March 18, 1982).

"Amine-Enhanced Photodegradation of Polychlorinated Biphenyls, Final Report," DE82021378, Battelle Pacific Northwest Laboratories, Richland, Washington; National Technical Information Service, Springfield, Virginia.

Anon., "EPA Hazardous Waste Site Ranking Blasted," *Chem. Eng. News,* 6 (November 6, 1981).

Baasel, W. D., et al., "Multimedia," *Environ. Goals,* 37–51 (October 1980).

Baker, R. A., "Taste and Odor in Water: A Critical Review," Final Report F-A2333, Manufacturing Chemists Association, Washington, D.C. 1978.

Bretherick, L., *Handbook of Reactive Chemical Hazards,* 2nd ed., Butterworth, London, 1979.

Carlson, G. A., and Collin, R. L., "Toxic Materials in the Environment," Technical Paper No. 52, Department of Environmental Conservation, Albany, N.Y., June 1978.

Cumberland, R. F., "The Control of Hazardous Chemical Spills in the United Kingdom," *J. Hazard. Mater.* **6,** No. 3, 277–287 (1982).

Fredette, P. E., Schultz, T. J., and Dominiak, B. W., "Chemical Suppression of Air Pollution During the Thermal Disposal of Hazardous Wastes," *Hazard. Mater.* **6,** No. 4, 383–390 (September 1982).

Gibbs, Lois M., *Love Canal: My Story,* State Univ. of New York Press, Albany, 1982.

Giffin, C. E., and Jahnsen, V. J. "Chemical Analysis of Waste Sites Using GC/MS Instrumentation," presented at the American Chemical Society Meeting, Las Vegas, Nevada, August 29, 1980.

Hilbert, J. M., and Cohen, A., "Disposal of Toxic Waste," *Chem. Eng. News,* 3 and 37 (September 5, 1983).

"Initial Emission Assessment of Hazardous Waste Incineration Facilities," DE82017507, Oak Ridge National Laboratory, National Technical Information Service, Springfield, Virginia.

"Interim Status Standards for Owners and Operators of Hazardous Waste, Treatment, Storage, and Disposal Facilities," *Fed. Reg.,* **46,** No. 221, 56592–56596 (November 17, 1981). (U.S. EPA rule on "Lab-Pacs.")

Jones, C. J., and McGugan, P. J. "An Investigation of the Evaporation of Some Volatile Solvents from Domestic Waste," *J. Hazard. Mater.* **2,** No. 3, 235–252 (August 1978).

Jones, C. J., "The Leaching of Some Halogenated Organic Compounds from Domestic Waste," *J. Hazard. Mater.* **2,** No. 3, 227–234 (August 1978).

Lazar, E. C., "Damage Incidents from Improper Land Disposal," *J. Hazard. Mater.* **1,** No. 2, 157–164 (January 1976).

"Liability Coverage: Requirements for Owners or Operators of Hazardous Waste Treatment, Storage, and Disposal Facilities," PB 83-144675, ICF Inc., Washington, D.C.; National Technical Information Service, Springfield, Virginia.

Metry, A. A., *The Handbook of Hazardous Waste Management,* Technomic Publishing, Westport, Conn., 1980.

Meyer, E., *Chemistry of Hazardous Materials,* Prentice-Hall, Englewood Cliffs, N.J. 1976.

Mintz, M., "Jail Terms Sought for Business Health, Environment Violators," *Washington Post,* A-1, A-13 (November 25, 1979).

Patrick, P. K., "Treatment and Disposal of Hazardous Wastes in Western Europe," *J. Hazard. Mater.* **1,** No. 1, 45–58 (September 1975).

"Pentagon's Waste Sites Represent Huge, Neglected Crisis, Critics Say," *Wall Street Journal,* July 22, 1983, p. 19. ("We can't afford any longer to condone a double standard that shields federal waste sites from strict enforcement," says an official of the National Wildlife Federation.)

Piasecki, B., "Unfouling the Nest," *Science 83* **4,** No. 7, 76–81 (September 1983). (Explores alternative methods of disposal or destruction vs. landfills.)

Pipitone, D., *Safe Storage of Laboratory Chemicals,* Wiley–Interscience, New York, 1984.

"Poison Problem: Fearing New Love Canal, Chemical Firms Stress Safer Disposal of Hazardous Waste," *Wall Street Journal,* June 30, 1983, p. 58.

Pojasek, R. B., *Toxic and Hazardous Waste Disposal (Vol. 2) Options for Stabilization–Solidification,* Ann Arbor Science Publishers, Ann Arbor, 1979.

Powers, P. W., *How to Dispose of Toxic Substances and Industrial Wastes,* Noyes Data Corp., Park Ridge, N.J. 1976.

Proceedings and papers, *International Symposium on Industrial and Hazardous Waste,* June 24, 1985–June 27, 1985, Alexandria, Egypt (joint sponsorship of Alexandria University, the Egyptian Academy of Scientific Research and Technology, and the ASTM Committee D-34 on Waste Disposal). (Contact Kathy Greene, ASTM Publications, 1916 Race Street, Philadelphia, PA 19103, or Richard Conway, Union Carbide Corporation, P.O. Box 8361, South Charleston, WV 25303.)

Proceedings of the Ninth Annual Research Symposium on Land Disposal, Incineration and Treatment of Hazardous Waste, May 2, 1983–May 4, 1983, Ft. Mitchell, Kentucky. (Contact Robert Landreth, U.S. EPA, Solid and Hazardous Waste Research Division, Municipal Environmental Research Laboratory, Cincinnati, OH 45268.)

Proceedings of the U.S. EPA National Conference on Management of Uncontrolled Hazardous Waste Sites, October 15, 1980–October 17, 1980, Washington, D.C.

Roland, R. A., "Chemist or Culprit: A Look Ahead," *The Chemist of AIC*, 4, 18 (August/September 1980).

Roulier, M. H., Landreth, R. E., and Carnes, R. A., "Current Research on the Disposal of Hazardous Wastes," *J. Hazard. Mater.* 1, No. 1, 59–64 (September 1975).

Senkan, S. M., and Stauffer, N. W., "What To Do With Hazardous Waste," *Technology Review (M.I.T.)* 84, No. 2 (November/December 1981).

Silvestri, A., "Development of a Kit for Detecting Hazardous Material Spills in Waterways," EPA–IAG-0546, EPA-600/2-78-055, Edison, N.J., March 1978.

Thibodeaux, L. J. *Chemodynamics,* Wiley–Interscience, New York, 1979.

Thibodeaux, L. J., "Estimating the Air Emissions of Chemicals from Hazardous Waste Landfills," *J. Hazard. Mater.* 4, No. 3, 235–244 (January 1981).

Thibodeaux, L. J., "Estimating the Air Emissions of Chemicals from Hazardous Waste Landfills," *J. Hazard. Mater.* 4, No. 3, 235–244 (January 1981).

"Those Toxic Chemical Wastes," *Time,* 58–69 (September 22, 1980).

Touhill, C. J., "Guidelines for Developing a Hazardous Waste Management Plan," *Plant Eng.* 37, No. 16, 81–83 (August 4, 1983).

Touhill, C. J., Shuckrow, A. J., and Pajak, A. P., "Hazardous Waste Management at Abandoned Dump Sites—Evolving Perspectives," *J. Hazard. Mater.,* 6(3), 261–265 (1982).

Whitmore, F. C., and Carnes, R. A., "Windmills, Incineration and Siting," *J. Hazard. Mater.* 5, No. 1–2, 103–109 (October 1981).

Wilson, D. C., *Waste Management Planning, Evaluation, Technologies,* Oxford Univ. Press, Oxford, U.K. 1981.

Zajic, J. E., and Himmelman, W. A., *Highly Hazardous Material Spills and Engineering Emergency Plans,* Marcel Dekker, New York 1978.

Monitoring Systems

"Agent-Orange Issue is Colored by Need for More Data" (Letters to the Editor) *Wall Street Journal,* September 1, 1983, p. 25.

Faust, E. A., "Monitoring of Ethylene Oxide and Nitrous Oxide in Hospital Environments," *Ind. Hyg. News* 6, No. 5, 45–47 (July 1983).

Goodenough, M., Mathieu, R. J., and Lawson, A. E., "Gas Chromatography Applications in Industrial Hygiene Analyses," *Ind. Hyg. News* 6, No. 5, 40–45 (July 1983).

Hazardous Wastes Handbook, 4th Ed. (422 pages), Government Institutes, Inc., Department PE, P.O. Box 1096, Rockville, MD 20850.

Krause, L. A., "Industrial Hygiene Chemistry—Who Needs It?" *Ind. Hyg. News* 6, No. 5, 34 (July 1983).

Miller, H. A., "Monitor Protects Workers. Against Exposure to Low-Level Toxic Gases," *Ind. Hyg. News* 6, No. 5, 47–48 (July 1983). (Describes arsine and phosphine monitor for routine use.)

Phillips, C., "Morale, Enforcement in EPA Office Improves Since Burford Resignation," *Wall Street Journal,* July 28, 1983, pp. 21 and 28.

Shields, E. J., "Appraising Wastewater Discharge Compliance Levels," *Pollut. Eng.* XV, No. 7, 33–37 (July 1983).

Superfund Notebook, 3rd Ed. (233 pages), 1983, Government Institutes, Inc., Department PE, P.O. Box 1096, Rockville, MD 20850.

"U.S. Scientists Plan to Step Up Research on Effects of Dioxin," *Wall Street Journal,* July 29, 1983, p. 6.

Winslow, R., "Hunting for Dioxin Exposes Field Teams to Array of Hazards: Garbed Like Spacemen, They Face Toxic Risks, Heat; 'We've Got the Best Place'," *Wall Street Journal,* July 26, 1983, pp. 1 and 8.

RCRA-SUPERFUND

Baasel, W. D., et al, "Multimedia Environmental Goals," *CEP.,* October 1980, pp. 37–51.

Baker, R. A., Taste and Odor in Water—A Critical Review, Final Report, P-A2333, Manu. Chemists Association (now CMA), Washington, D.C. 1978.

Carlson, G. A., and Collin, R. L., *Toxic Materials in the Environment,* Technical Paper No. 52, Department of Environmental Conservation, Albany, New York, June 1978.

Giffin, C. E., and Jahnsen, V. J., *Chemical Analysis of Waste Sites Using GC/MS Instrumentation,* August 1980, American Chemical Society, Las Vegas, Nevada.

Roland, R. A., "Chemist or Culprit: A Look Ahead," *The Chemist of AIC.,* August/September 1980, pp. 4 and 18.

Taylor, R. E., U.S. Files 5 Suits Under Superfund Over Waste Sites, *Wall Street Journal,* December 14, 1983, p. 7.

"Those Toxic Chemical Wastes," *TIME,* Sept. 22, 1980, pp. 58–69.

Silvestri, A., *Development of a Kit for Detecting Hazardous Material Spills in Waterways,* EPA-IAG-0546, EPA-600/2-78-055, Edison, New Jersey, March 1978.

Treatment of Waste

"Access to Employee Medical Records," OSHA, *Fed. Reg.* **47,** 8349 (February 26, 1982).

Anon., "DOD Tied to Illegal Hazardous Waste Disposal," *Chem. Eng. News,* 7 (August 22, 1983) (question of illegal disposal of thousands of gallons of PCB-containing contaminated oil).

Argonne National Laboratory, "Hazardous Waste and Materials Management Plan: A Case Study," DE83002296, Argonne National Laboratory, Illinois. (available from the NTIS).

Audubon Naturalist Society, *"A Household Guide to Hazardous Wastes,"* Insert 1 to *Audubon Naturalist News,* July–August 1983, Audubon Naturalist Society, 8900 Jones Mill Road, Chevy Chase, MD 20815. (Presents information on safe disposal and substitution of hazardous household chemicals.)

Bendersky, D., *Resource Recovery Processing Equipment,* Noyes Data Corp., Park Ridge, N.J., 1982.

Berkowitz, J. B., *Unit Operations for Treatment of Hazardous Industrial Wastes,* et al., Noyes Data Corp., Park Ridge, N.J., 1978.

Berry, E. E., and MacDonald, L. P., "Experimental Burning of Used Automobile Crankcase Oil in a Dry-Process Cement Kiln," *J. Hazard. Mater.* **1,** No. 2, 137–156 (January 1976).

Biodegradation Techniques for Industrial Organic Wastes, D. J. DeRenzo (Ed.), Pollution Technology Review No. 65, Noyes Data Corp., Park Ridge, N.J., 1980.

Bonner, T., *Hazardous Waste Incineration Engineering,* Noyes Data Corp., Park Ridge, N.J., 1981.

Bowders, J. J., Koerner, R. M., and Lord, A. E. Jr., "Buried Container Detection Using Ground Probing Radar," *J. Hazard. Mater.* **7,** No. 1, 1–18 (Nobember 1982).

Buchel, K. H., *The Chemistry of Pesticides,* Wiley, New York, 1982.

"Bursting the Bubble" (Editorial), *Wall Street Journal,* September 1, 1982, p. 14 (discusses recent decisions on the Clean Air Act).

Cheremisenoff, N. P., *Industrial and Hazardous Waste Impoundment,* Ann Arbor Science Publishers, Ann Arbor, 1979.

abstracts (in English) of documents in various languages providing *CIS* information on rules, regulations, and practices contributing to safety and health in laboratory work. Information Search, Retrieval Service of the International Occupational Safety and Health Information Centre, Bureau of International du Travail, CH-1211, Geneve 22, Switzerland, 1982.

Corbin, M. H., and Lederman, P. B., "Hazardous Landfill Management, Control Options," *J. Hazard. Mater.* **5,** No. 3, 235–254 (February 1982).

"Current Focus—A Hazardous Waste Primer," Publication No. 402, League of Women Voters of the United States, 1730 M Street, NW, Washington, D.C. 20036 (8 pp) 1980.

Debbrecht, F. J., and Neel, E. M., "Instruments for Detecting Fugitive Emissions—Types, Operating Principles and Selection Guidelines," *Plant. Eng.* **37,** No. 17, 54–57 (August 18, 1983). (Discusses instrumentation useful in detecting and measuring fugitive emissions from plant equipment processing organic substances potentially damaging to human health and the environment.)

DeRenzo, D. J., *Unit Operations for Treatment of Hazardous Industrial Waste,* Noyes Data Corp., Park Ridge, N.J., 1978.

"Development Document for Expanded Best Practicable Control Technology," "Best Conventional Pollutant Control Technology," "Best Available Technology," "New Source Performance Technology," and "Pretreatment Technology in the Pesticide Chemicals Industry," EPA PB83-153171. (Available from the NTIS, "Asbestos Removal and Disposal Information System: A User's Guide," DE83001704, Oak Ridge National Laboratory, Tennessee.)

"Dow Chemical Fights Effect of Public Outcry over Dioxin Pollution," *Wall Street Journal,* June 28, 1983, pp. 1 and 20. (Evidence shows it told industry of the dangers before agent orange sale.)

Dyer, J. C., and Mignone, N. A., *Handbook of Industrial Residues,* Noyes Data Corp., Park Ridge, N.J., 1983.

Edwards, B. H., Paullin, J. N., Coghlan-Jordan, K., *Emerging Technologies for the Control of Hazardous Wastes,* Noyes Data Corp., Park Ridge, N.J., 1983.

"Empty Hazardous Waste Container, Final Rule," *Fed. Reg.* **47,** 36092 (August 18, 1982).

"EPA,DOD Sign Superfund Agreement," *Chem. Eng. News,* 14 (August 22, 1983). [DOD will facilitate hazardous-waste clean-up at 265 DOD facilities for storing, treating, or disposing of hazardous wastes (83% of all federal hazardous-waste operations). Executive Order 12316 makes DOD responsible for clean-up.]

"EPA Study Finds Chesapeake Bay Seriously Threatened by Pollution," *Washington Post,* March 28, 1983, p. D5. (Growing accumulations of toxic substances as well as chemicals from agricultural fertilizers blamed for pollution by Chesapeake Bay Foundation.)

"EPA Updates Efforts for Hazardous Wastes," *Chem. Eng. News,* 7 (September 5, 1983).

"Exemption of Chemical Substances from Premanufacture Notification Under TSCA," *Fed. Reg.* **47,** 33896 (August 4, 1982).

"Exemption of Polymers of Low-Risk from Premanufacture Notification," *Fed. Reg.* **47,** 33924 (August 4, 1982).

"Final Rule Under FIFRA Regulations Regarding Certain Organisms," *Fed. Reg.* **47,** 23928 (June 2, 1982).

Forums on Hazardous Waste Management at Academic Institutions, ACS Northeast Regional Meeting, October 1981 and ACS MARM meeting, April 1982, Department of Public Affairs, American Chemical Society, Washington, D.C., 1983.

Geological Aspects of Industrial Waste Disposal, Ad Hoc Committee, National Research Council, Washington, D.C., 1982.

Hazardous Waste Bulletin (a periodical), 177 Greenwich Avenue, Box 1274, Stamford, CT 06904.

"Hazardous Waste Disposal: The Issues and the Controversy," PB82-177114, Illinois State Office of the Governor, Springfield, Illinois.

"Hazardous Materials Response Project: Program Plan," PB82-193046, NOAA, Boulder, Colorado.

Hooper, G. V., *Offshore Shipping and Platform Incineration of Hazardous Wastes,* Noyes Data Corp., Park Ridge, N.J., 1981.

How to Dispose of Oil and Hazardous Chemical Spill Debris, A. Breull (Ed.), Pollution Technology Review No. 87, Noyes Data Corp., Park Ridge, N.J., 1981.

"Illegal Toxic Dumping is Laid to Organized Crime," *New York Times,* June 5, 1983, p. 1.

Janson, D., "Jersey Aides Calm About Heading Waste Site List," *New York Times,* September 2, 1983, p. A-17. (New Jersey has 85 of the nation's 546 most dangerous sites.)

Jones, C. J., McGugon, P. J., Smith, A. J. and Wright, W. J., "Adsorption of Some Toxic Substances by Waste Components," *J. Hazard. Mater.* **2,** No. 3, 219–226 (August 1978).

Katerman, G., and Pijpers, F. W., *Quality Control in Analytical Chemistry,* Wiley, New York, 1981.

Kirk-Othmer, *Encyclopedia of Chemical Technology,* new 3rd ed. (in 25 vols.), Wiley, New York, 1983–1984.

Klumpp, G. W., *Reactivity in Organic Chemistry,* Wiley, New York, 1982.

"Land Disposal of Hazardous Waste: Proceedings of the Annual Research Symposium (8th)," PB82-173022, March 8, 1982–March 10, 1982, Ft. Mitchell, Kentucky Southwest Research Institute, San Antonio.

Laughlin, R. G. W., Gallo, T., and Robey, H., "Wet Air Oxidation for Hazardous Waste Control," *J. Hazard. Mater.* **8,** No. 1, 1–10 (June 1983).

Lord, A. E., "The Identification and Location of Buried Containers Via Non-Destructive Testing Methods," *J. Hazard. Mater.* **5,** No. 3, 221–234 (February 1982).

Louis, E. T., "In Danville, N.H., Mr. Hunt Tries to Get Rid of 15 Million Old Tires," *Wall Street Journal,* August 25, 1983, p. 21.

Lowrance, W. W., *Assessment of Health Effects at Chemical Disposal Sites,* Proceedings of a symposium, June 1, 1981–June 2, 1981, William Kaufmann, Inc., Los Altos, CA 94022.

"Manville Holders Face a Dilemma: Sell at Loss or Gamble on Outcome," *Wall Street Journal,* September 3, 1982, p. 17. (See also "Canadian Asbestos Firms Blame Problems Mainly on Recession, Rather than Suits," in same issue on p. 5.)

"Manville's Chapter 11 Filing Spotlights Tiny Firm That Probes Health Hazards," *Wall Street Journal,* August 31, 1982, p. 6.

Martin, A. E., *Small-Scale Resource Recovery Systems,* Noyes Data Corp., Park Ridge, N.J., 1982.

McKinney, J. D., Albro, P. W., Cox, R. H., Hass, J. R., and Walters, D. B., "Chemical Epidemiology (and Analytical Chemistry)" in *The Pesticide Chemist and Modern Toxicology,* S. K. Bandal et al. (Eds.), ACS Symposium Series No. 160, American Chemical Society, Washington, D.C., 1981, Chap. 25, pp. 448–460.

Metry, A. A., *The Handbook of Hazardous Waste Management,* Technomic Publishing Co., Westport, Conn, 1980.

Muller, B. W., Brodd, A. R., and Leo, J. P., "Hazardous Waste Remedial Action—Picillo Farm, Coventry, R.I., An Overview," *J. Hazard. Mater.* **7,** No. 2, 113–130 (January 1983).

Offshore Ship and Platform Incineration of Hazardous Wastes, G. V. Hooper (Ed.), Pollution Technology Review No. 79, Noyes Data Corp., Park Ridge, N.J., 1981.

Oil Spill Cleanup and Protection Techniques for Shorelines and Marshlands, A Breuel (Ed.), Pollution Technology Review No. 78, Noyes Data Corp., Park Ridge, N.J., 1981.

Parr, J. F., *Land Treatment of Hazardous Wastes,* Noyes Data Corp., Park Ridge, N.J., 1983.

Pesticide Disposal and Detoxification: Processes and Techniques, Pollution Technology Review No. 81, Noyes Data Corp., Park Ridge, N.J., 1981.

Plambeck, J. A., *Electroanalytical Chemistry: Basic Principles and Applications,* Wiley, New York, 1982.

"Poison Pangs?" (Review and Outlook Editorial), *Wall Street Journal,* August 31, 1982, p. 24 (refers to Soviet use of chemical warfare).

"Problems Have Long Plagued Asbestos Firms," *Wall Street Journal,* August 30, 1982, p. 15. (See also "The Price of Toxic Torts" in the same issue on p. 12.)

Proceedings of the 1982 Hazardous Materials Spill Conference, April 1982, Government Industries, Rockville, Maryland, 1982 (includes papers dealing with hazardous-waste disposal sites).

Publications available from *Government Institutes, Inc.,* 966 Hungerford Drive, #24, Rockville, MD 20850:

Environmental Law Handbook, 7th ed., 1983, 528 pp., 1983.

Superfund Notebook, 3rd ed., 233 pp., 1983.

Hazardous Material Spills, 510 pp., April 1982.

EPA Guidebook, 245 pp., 1983.

Hazardous Wastes Handbook, 4th ed., 656 pp., 1983.

EPA's RCRA Inspection Manual, 300 pp., 1983.

Environmental Glossary, 303 pp., 1983.

EPA's TSCA Inspection Manual, 300 pp., 1983.

Incineration Systems Course Notebook, 411 pp., 1983.

Quinlivan, S. C., et al., "Sources, Characteristics and Treatment and Disposal of Industrial Wastes Containing Hexachlorobenzene," *J. Hazard. Mater.* **1,** No. 4, 343–360 (March 1977).

"Radical Tactic: Manville's Big Concern As It Files in Chapter 11 is Litigation, Not Debt," *Wall Street Journal,* August 27, 1982, p. 1. (See also full-page ad: "Despite Strong Business, Litigation Forces Manville to File for Reorganization," *Wall Strett Journal,* August 27, 1982, p. 29. Action refers to 16,500 lawsuits related to health effects of asbestos, with many more lawsuits projected.)

"Racketeering Charges Due in Penna. Waste-Dumping Case," *Wall Street Journal,* July 8, 1983, p. 8. (Grand jury recommends racketeering charges against man who allegedly dumped 3.6 million gallons of industrial waste into an abandoned coal mine under Scranton, and illegally buried 9721 drums of waste at the Scranton landfill.)

RCRA and the Laboratory, Department of Public Affairs, American Chemical Society, Washington, D.C., 1983.

Requirements and Alternatives for a Management Information System to Support Hazardous Waste Enforcement, by American University, Washington, D.C., 1983.

Rice, R. G., and Browning, M. F., *Ozone Treatment of Industrial Wastewater,* Noyes Data Corp., Park Ridge, N.J., 1981.

Rogers, H. W., Shaff, R. E., and Mullican, M. R., "Problems of Waste Chemical Handling at a Large Biomedical Research Facility," *J. Hazard. Mater.* **4,** No. 2, 191–206 (September 1980).

Rogoshewski, P., Bryson, H., and Wagner, K., *Remedial Action Technology for Waste Disposal Sites,* Noyes Data Corp., Park Ridge, N.J., 1983.

Rosen, M. J., *Surfactants and Interfacial Phenomena,* Wiley, New York, 1978.

Shabecoff, P., "E.P.A. Adds 133 Sites to Hazardous Waste List," *New York Times,* September 2, 1983, p. A-18. (Total now is 546.)

Shriner, R. L., Fuson, R. C., Curtin, D. Y., and Morrill, T. C., *Systematic Identification of Organic Compounds,* 6th ed., Wiley, New York, 1980.

Shuckrow, A. J., *Hazardous Waste Leachate Management Manual,* Noyes Data Corp., Park Ridge, N.J., 1982.

Shuckrow, A. J., and Pajak, A. P., "Bench Scale Treatability of Contaminated Groundwater at the Ott/Story Site, Part 1," *J. Hazard. Mater.* **7,** No. 1, 37–50 (November 1982).

Sittig, M., *Incineration of Industrial Hazardous Wastes and Sludges,* Noyes Data Corp., Park Ridge, N.J., 1979.

Sittig, M., *Landfill Disposal of Hazardous Wastes and Sludges,* Noyes Data Corp., Park Ridge, N.J., 1979.

Sittig, M., *Metal and Inorganic Waste Reclaiming Encyclopedia,* Noyes Data Corp., Park Ridge, N.J., 1980.

Sittig, M., *Organic and Polymer Waste Reclaiming Encyclopedia,* Pollution Technology Review No. 73, Chemical Technology Review 180 Noyes Data Corp., Park Ridge, N.J., 1981.

Small-Scale Resource Recovery Systems, A. E. Martin (Ed.), Pollution Technology Review No. 89, Noyes Data Corp., Park Ridge, N.J., 1982.

Stoner, D. L., Smathers, J. B., Duncan, D. D., Clapp, D. E., and Hyman, W. A., "Engineering a Safe Hospital Environment," in *Waste Disposal,* Wiley–Interscience, N.Y., 1982, pp. 69–76.

Survey of Laboratory Practices and Policies for Employee Protection from Exposure to Chemicals, Department of Public Affairs, American Chemical Society, Washington, D.C., 1983.

Tenzer, R., Ford, B. Jr., Mattox, W., and Brugger, J. E., "Characteristics of the Mobile Field Use System for the Detoxification/Incineration of Residuals from Oil and Hazardous Materials Spill Clean-Up Operation," *J. Hazard. Mater.* **3,** No. 1, 61–76 (February 1979).

Thibodeaux, L. J., et al., "Models of Mechanisms for the Vapor Phase Emission of Hazardous Chemicals from Landfills," *J. Hazard. Mater.* **7,** No. 1, 63–74 (November 1982).

Thorsen, J. W., "Bruin Lagoon: Remedial Cleanup of Hazardous Waste Sites Under Superfund," Publication No. I-1190, presented at the 14th Mid-Atlantic Industrial Waste Conference, June 1982, University of Maryland, College Park, Maryland; Roy F. Weston, Inc., West Chester, PA 19380.

"Toxic City: Iowa Town Learns to Live with Chemical Waste; Economic Woes Overshadow Pollution Worries," *Wall Street Journal,* April 27, 1983, p. 60. (Charles City concerned about 6 million pounds of toxic chemicals in 8 and one-half-acre closed dump site.)

"Toxic Substances Control Act: Report to Congress for FY 81," PB82-195330, U.S. EPA.

Tyagi, S., Lord, A. E., Jr., Koerner, R. M. "Use of a Metal Detector to Detect Buried Drums in Sandy Soil," *J. Hazard. Mater.* **7,** No. 4, 375–382 (March 1983).

Tyagi, S., Lord, A. E., Jr., and Koerner, R. M., "Use of a Very Low Frequency Electromagnetic Method at 9.5 KHz to Detect Buried Drums in Sandy Soil," *J. Hazard. Mater.* **7,** No. 4, 353–374 (March 1983).

"Unproductive Liability" (Review and Outlook) *Wall Street Journal,* September 9, 1982, p. 26 (discusses product liability).

"Waste Disposal Site Survey," Publication No. 052-070-051-48-5, Superintendent of Documents, U.S. GPO, Washington, D.C., 1983.

8

Dioxin (TCDD), Dibenzofurans, and Related Compounds

Over 4 million molecules are known, but few have received the publicity in recent times that "dioxins" have attracted. This imprecise term has been frequently used, especially by the electronic and print media, to strike an emotional note frequently out of proportion to sober scientific reality. In an age when Chad, El Salvador, Lebanon, herpes, AIDS, MS, pot, coke, heroin, budget deficiencies, inflation, and cigarette smoking compete with other "horror stories" for the national nightly news, the dioxins are assumed to have strong interest, although most viewers, even those chemically trained, have very limited background knowledge of the alleged hazards in the latest fragmentary stories.

Often overlooked is the fact that dioxins are not deliberately manufactured molecules, but are formed as unwanted byproducts in the preparation of the trichlorophenols, PCBs, and certain herbicides which contain the specific atomic configuration and are subjected to unusual conditions (of temperature and pressure), see Fig. 8.1).

8.1. CHEMISTRY OF DIOXINS AND RELATED SUBSTANCES

The "dioxin" usually referred to by the media is one of 75 isomers with various degrees of chlorination and has the basic chemical structure

Dibenzo-p-dioxin

which when substituted becomes

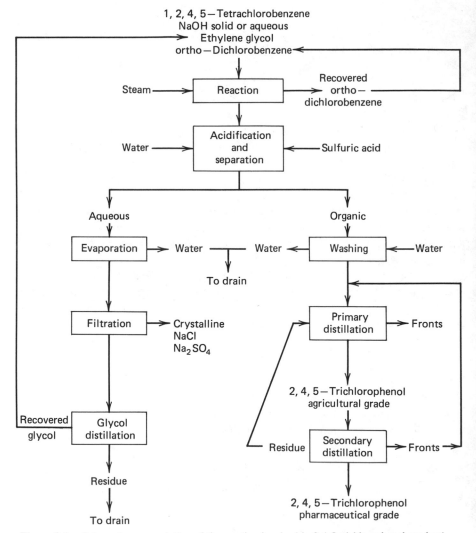

2,3,7,8-Tetrachloro-
dibenzo-p-dioxin (TCDD)

Figure 8.1.   Schematic representation of the reaction involved in 2,4,5-trichlorophenol production.

Other molecules closely associated with the dioxins are the dibenzofurans, which are formed by the condensation reaction involving tetrachlorobenzene and sodium trichlorophenate:

2,3,7,8-Tetrachlorodibenzofuran (TCDF)

These molecules, and the other isomers, are formed in the manufacture of 2,4,5-trichlorophenol:

The above reactions are essential to the production of the herbicide 2,4,5-T, which has received much attention recently because of its role in the Agent Orange questions, as well as being a useful herbicide in its own right:

2,4,5-Trichlorophenol   Sodium salt of 2,4,5-T   2,4,5-T

(see Figs. 8.2–8.4).

Another source of both dioxins and polychlorinated dibenzofurans (PCDFs) is the manufacture of commercial-grade pentachlorophenol.

The polychlorinated biphenyls (PCBs) are excellent heat-transfer agents, and were widely used in electrical transformers and capacitors, until banned in the United States in 1977. PCBs were considered relatively innocuous materials in the United States and Europe until 1968 when people in Japan became ill from cooking-rice oil contaminated with PCBs and PCDFs. The incident occurred in the Fukuoka prefect in 1968, and received much attention since chloracne, a distinctive skin disease, was observed in many persons. Persons involved were identified as having Yusho (oil) disease. The contamination apparently occurred when PCBs (Kanechlor 400), used to heat the oil under reduced pressure, leaked into the oil and reactions occurred producing the dioxins, furans, and other impurities. PCBs have not always been disposed of safely (see Fig. 8.5).

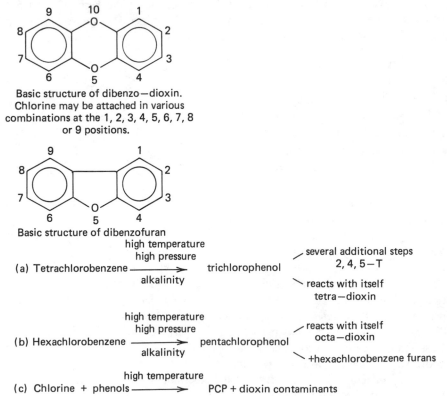

Sodium trichlorophenate   TCDD

Figure 8.2.   Formation of TCDD by Condensation of Sodium Trichlorophenate.

Depending on the reaction rate, concentrations, and temperature control, various combinations of the dioxins and dibenzofurans are formed. Should the temperature rise in the initial reaction stages, a condensation reaction involving tetrachlorobenzene and 2,4,5-sodium trichlorophenate would occur and 2,3,7,8-TCDF would be formed. Professor Rappe of Sweden feels that an accident during this reaction at an early stage favors PCDF formation, and at the end of the reaction favors TCDD production. At Seveso, Italy, the accident occurred 6 and one-half hr after the completion of the run; the soil analysis confirms that 2,3,7,8-TCDD was the major contaminant present.

More details on the chemistry of the compounds will be found in the references at the end of this chapter.

Basic structure of dibenzo–dioxin.
Chlorine may be attached in various
combinations at the 1, 2, 3, 4, 5, 6, 7, 8
or 9 positions.

Basic structure of dibenzofuran

(a) Tetrachlorobenzene $\xrightarrow[\text{alkalinity}]{\substack{\text{high temperature}\\ \text{high pressure}}}$ trichlorophenol $\nearrow$ several additional steps 2, 4, 5–T
$\searrow$ reacts with itself tetra–dioxin

(b) Hexachlorobenzene $\xrightarrow[\text{alkalinity}]{\substack{\text{high temperature}\\ \text{high pressure}}}$ pentachlorophenol $\nearrow$ reacts with itself octa–dioxin
$\searrow$ +hexachlorobenzene furans

(c) Chlorine + phenols $\xrightarrow{\text{high temperature}}$ PCP + dioxin contaminants

Figure 8.3.   Formation of dioxins and furans.

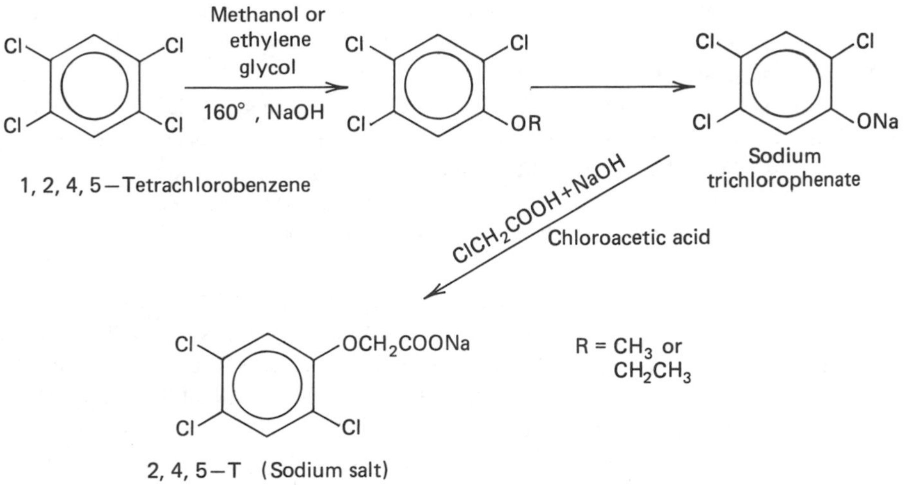

Cl — [ring] — Cl
Cl — [ring] — Cl

1, 2, 4, 5 — Tetrachlorobenzene

Methanol or ethylene glycol

160° , NaOH

→

Cl — [ring] — Cl
Cl — [ring] — OR

→

Cl — [ring] — Cl
Cl — [ring] — ONa

Sodium trichlorophenate

ClCH₂COOH + NaOH
Chloroacetic acid

Cl — [ring] — OCH₂COONa
Cl — [ring] — Cl

2, 4, 5 — T (Sodium salt)

R = CH₃ or CH₂CH₃

Figure 8.4. General Scheme for Production of 2,4,5-T.

8.2. TOXICITY AND HUMAN EXPOSURES TO DIOXINS AND RELATED MOLECULES

As noted previously (page 48), not all animal species or all humans react in the same way to a given exposure or dose of a chemical or drug.

The misleading statement that dioxin (TCDD) is "one of the deadliest substances known" or that "it is one of the deadliest manmade substance" is based on its extreme toxicity to guinea pigs. As little as 0.6 μg/kg of body weight orally of

Figure 8.5. Actual roadside sign in North Carolina where 210 miles of highways were used as disposal sites for PCBs. Courtesy of U.S. EPA.

this substance will kill half the male guinea pigs receiving the dose (known as the oral LD_{50}). Illness occurs immediately and death occurs within a week or 10 days after the dose is administered.

Dioxin (TCDD) is much less toxic to mice than to guinea pigs. The guinea pig is 500 to 10,000 times more sensitive to TCDD than is the hamster (which is the least sensitive animal tested). Rabbits, mice, and monkeys are roughly 200 times less sensitive than guinea pigs and 50 times more sensitive than hamsters. (Such comparisons of relatively toxicity must be used with great care, as noted on page 56).

The single dose LD_{50} for monkeys for TCDD is 50 μg/kg of body weight. Human doses are considered in the same range of 50–100 μg/kg of body weight. Using the figure of 50 μg/kg, TCDD is about 10 times as toxic as hydrogen cyanide, but such *comparisons must be used with great care,* since the ability of HCN to enter the body is many times greater than TCDDs.

The single doses LD_{50} for several substances, for comparison, are:

Dioxin (TCDD)	50 μg/kg
HCN	500 μg/kg
Cyanide salts	2000 μg/kg
Cantharidin	300 μg/kg
Nicotine	600 μg/kg
Colchicine	300 μg/kg
Digitoxin (digitalis)	35 μg/kg

Another tabulation of comparative toxicity follows:

		Minimum Lethal Dose (Moles/Kg of Body Weight)
Botullinum toxic A	Mouse	3.3×10^{-17}
Tetanus toxin	Mouse	1.0×10^{-15}
Diphtheria toxin	Mouse	4.2×10^{-12}
2,3,7,8-TCDD	Guinea pig	3.1×10^{-9}
Bufotoxin	Cat	5.2×10^{-7}
Curare	Mouse	7.2×10^{-7}
Strychnine	Mouse	1.5×10^{-6}
Muscarin	Cat	5.2×10^{-6}
Di-isopropylfluoro phosphate	Mouse	1.6×10^{-5}
Sodium cyanide	Mouse	2.0×10^{-4}

Source. EPA, as published in *Chem. Eng. News*, 45 (June 6, 1983).

Dioxin (TCDD) is stable to heat, acids, and alkali. Its solubility in water is very low (2×10^{-4} ppm). It is slightly soluble in fats (44 ppm in lard oil); 570 ppm of dioxin dissolves in benzene and 1400 ppm in 0-dichlorobenzene. In soil, where the soil microbes degrade or decompose the substance, the half-life of TCDD in Utah and Florida is 320 and 230 days, respectively. The dioxin half-life at Severso, where soil conditions and bacteria are different, may be 2 or 3 years, or possibly longer.

At temperatures of 700°C (1292°F), only 50% TCDD is decomposed in 21 sec, while at 800°C (1472°F) the decomposition is reported complete. A residence time of at least 2 sec in an incinerator has been required in such operations. Considerable uncertainty exists in the literature as to the TCDD formation (if any) from burning of vegetation sprayed with 2,4,5-T. In forest fires, where the temperature of 1200°C (2192°F) is known, the dioxins probably are completely, or nearly completely, destroyed. Pyrolysis of technical-grade PCB mixtures will yield some 30 major and 30 minor polychlorinated dibenzofurans.

As will be noted later in this chapter in the description of the decontamination difficulties of the Binghamton State Office Building, when PCBs are directly involved in fire, furans, including 2,3,7,8-tetrachlorodibenzofuran or TCDF—the most toxic furans, as well as the dioxins, are found. Uncontrolled burning of the PCBs can be an important environmental source of these compounds, a point made by Professor Rappe of Sweden and by Dow Chemical Company in the United States. Dr. Otto Hutzinger of the University of Amsterdam has found that, upon burning, lignin and PVC produce dioxins. The State of New York is considering regulating the use of products that generate smoke with unacceptably high toxic potency. (See A. D. Little, *Study to Access the Feasibility of Incorporating Combustion Toxicity Requirements into Building Material and Furnishing Codes of N.Y. State*, Ref. 88712, Cambridge, Mass., May 1983). A recent incident in the garment district of Manhattan involved 150 gal of transformer oil containing PCBs in an electrical vault three stories under the street, but the utility involved claims that no health hazard was presented to the fire fighters. Considerable smoke and soot were observed. (See K. A. Hudges, "Water-Main Break in Manhattan Leads to Power Outage," *Wall Street Journal,* August 11, 1983, p. 28.)

PCB IN PERSPECTIVE

Considerable additional study has been given the potential and actual health hazards of commercial grade PCB as used in the U.S. electrical industry, including studies of persons exposed for many years in the handling and filling of capacitors and transformers. Dr. Seymour Friess has summarized many of these findings: "At human exposure levels as high as those previously encountered in occupational settings, PCBs can cause chloracne and increases in one or more liver enzymes. Neither represents a serious or life-threatening problem. From present evidence, no firm associations can be made between PCB exposures and any other human health effects, even after long-term occupational exposures." See *PCB Perspectives*, booklet, National Electrical Manufacturers Association, 2101 L St., N.W., Washington, D.C. 20037, 1984, *Potential Health Effects in the Human from Exposure to Polychlorinated Biphenyls (PCBs) and Related Impurities*, prepared by Drill, Friess, Hays, Loomis & Shaffer, Inc., Arlington, VA 22209, February 12, 1982, and also *Literature Update on Potential Health Effects from Exposures to polychlorinated Biphenyls and Related Chlorinated Heterocycles*, prepared by V. A. Drill, S. L. Friess, H. W. Hays, and T. A. Loomis, Drill, Friess, Hays, Loomis & Shaffer, Inc., Arlington, VA 22209, December 20, 1983. See also *Highlights of the Potential Health Effects in the Human from exposure to Polychlorinated Biphenyls (PCBs) and Related Impurities*, booklet, National Electrical Manufacturers Association, 2101 L St., N.W., Washington, D.C. 20037, 1983.

Dioxins have also been noted in other fires. See R. R. Bumb, "Trace Chemistries of Fire: A Source of Chlorinated Dioxins," *Science, 210*, No. 4468, 385–390 (October 24, 1980).

Dioxin (TCDD) can be decomposed by ultraviolet light if dissolved in a suitable solvent such as methanol. Twenty-four hours were required for complete photolysis using artificial ultraviolet light; 36 hr were needed when using natural sunlight. Dioxin-containing substances, such as Agent Orange [a 1 : 1 mixture of butyl esters of 2,4-dichlorophenoxyacetic acid (2,4-D) and 2,4,5-T], can be chlorinated at high temperatures and pressures producing carbon tetrachloride. Oxidation by ruthenium tetraoxide (RuO_4) will also degrade TCDD.

8.3. HUMAN ASPECTS OF CLINICAL FINDINGS

The known or suspected effects of exposures to TCDD, based on studies and observations of the nearly 1000 persons who may have had exposures since 1949 in industrial accidents during the production of chlorinated phenols or 2,4,5-T (three of these 17 incidents were in the United States and may have involved a total of 317 persons), reveal:

1. *Chloracne.* This is an eruption of the skin causing blackheads, usually associated with small pale-yellow cysts. In severe cases, papulas or pustules may be observed. The major area involved is usually the face—under the eyes and behind the ears. It may persist for several years (15 or more after the last exposure). It should be noted parenthetically that chloracne in humans can be produced by exposure to:

    Chlornaphthalenes (CNs)
    Polychlorinated biphenyls (PCBs)
    Polybrominated biphenyls (PBBs)
    Polychlorinated dibenzo-p-dioxins (PCDD)
    Polychlorinated dibenzofurans (PCDFs)

3,4,3,4-tetrachloroazobenzene (TCAB)

3,4,3,4-tetrachloroazoxybenzene (TCAOB)

As noted in Fig. 2.4, the Dow Chemical Company has performed extensive research on the chloracne question.

2. *Enlarged Liver and Impairment of Liver Functions.* Increased transaminase values in serum over the normal may be found after exposure to TCDD, but this is not considered a reliable diagnostic tool. Clinical signs of liver dysfunction have been noted in up to 50% of persons with known TCDD exposures.

3. *Neurology.* Severe muscle pains aggravated by exertion, especially in the calves and thighs and chest area, and sensorial impairments of sight, hearing, smell, and taste have also been noted. In the central nervous system, lassitude, weakness, impotence, and loss of libido have been reported.

8.4. LONG-TERM EFFECTS OF TCDD-RELATED EXPOSURES

The following long-term effects have been observed:

1. *Immunocapability.* Persons exposed in the Coalite TCP explosion (United Kingdom in 1969), studied 10 years later, who still had chloracne, had some changes in immunological tests, but were in a comparable health state to other workers exposed without chloracne, and to an unexposed control group. Forty-five children exposed at the Seveso incident with maximum exposures of 235 μg/m² of soil (20 of whom had chloracne) had comparable levels of infective pathology, compared to 44 controls, 3 years later.

2. *Tetratogenic Effects.* Studies of several thousand women of childbearing age exposed to a known level of TCDD in Seveso have shown that the frequency of spontaneous abortions per number of calculated pregnancies and per number of women of childbearing age did not change either immediately after the accident or in the following 3 years compared to previous years. The claims made of abortions due to the spraying of 2,4,5-T in Oregon [as cited in the emotional book *A Bitter Fog* by Carol Van Strum (Sierra Club Books, San Francisco, May 1983)] have been largely discounted [see a review in *Chem. Eng. News,* 28–29 (August 15, 1983) and other independent sources]. Careless use of any chemical, as may have occurred in Oregon and in other areas, *cannot* be justified.

3. *Carcinogenic Aspects.* TCDD causes cancer in rats and mice; however, cancer mortality studies of five groups of exposed persons (410 workers) indicate no excess in mortality over the matching population 14 to 30 years after exposure. Increased stomach-cancer mortality was reported from the

71 workers in the BASF 1953 accident, but the significance of this study is in doubt.

In Sweden, frequency of soft-tissue sarcoma seems to be higher for railway workers spraying 2,4,5-T and other herbicides, but not in a similar group so engaged in Finland or in agricultural workers in New Zealand. The question is unresolved.

On July 29, 1983, a UPI press report quoted Philip Landrigan of the National Institute for Occupational Safety and Health as saying that an association has been found between dioxin and soft-tissue sarcoma, but no absolute proof is yet available. Soft-tissue sarcoma attacks muscles, fat, nerves, and/or connective fibers.

4. *Chromosome Aberrations.*   No abnormal chromosome aberrations have been reported in cytogenic studies in four exposed populations.

8.5. ACCEPTABLE DAILY INTAKE (ADI)

The acceptable daily intake (ADI) is the amount of TCDD which can be regularly or intermittently absorbed through all possible routes for the small concentrations absorbed for the entire life span via diet or other means of environmental contamination. In 1980 the Scientific Advisory Panel of the US EPA stated that a dose of 0.001 µg/kg of body weight per day of TCDD is, for all practical purposes, a no-observable-effect level including carcinogenic, teratogenic, and reproductive risk. The U.S. Food and Drug Administration recommended in 1981 that fish containing TCDD at a level of more than 50 ppt should not be consumed. The Center for Disease Control (CDC) recommends one part per billion (1 ppb) as an acceptable level of TCDD in soil. This is not intended as an all-purpose standard, but intended primarily for residential properties and other areas where children may ingest soil on a daily basis. Each contaminated site should be reviewed on a case-by-case basis. (The highest level of dioxin yet, found in Missouri, forced evacuation of three families near St. James. The EPA found levels up to 1800 ppb, six times higher than those initially found in Times Beach, Missouri.)

EPA has established criteria for 2,3,7,8 TCDD in surface water of 1.3×10^{-8} micrograms per liter [*Fed. Reg. 49,32* p. 5831 (Feb. 15, 1984)].

For 2,4,5-T, a joint committee of the Food and Agriculture Organization and the World Health Organization stated in 1980 that the "no-effect level" is 3 mg/kg per day, and the ADI for humans is 0.003 mg/kg per day when the TCDD content of the 2,4,5-T is 0.05 ppm.

The Seveso incident prompted the European Economic Community (EEC) to initiate studies on a directive aimed at preventing "the risk of incidents." The purpose of the directive—familiarly known as the "Seveso" directive—is the establishment of general criteria for the regulation of hazardous industrial activities in the 10 member countries.

8.6. REMEDIAL ACTION

Once it is determined that excessive levels of dioxins and/or other comparable hazardous materials are present in soil or ground areas, several approaches may be considered:

1. *Stabilization in Place.*   This involves sealing or otherwise covering the soil with a cover inside an impermeable barrier around the perimeter. A clay cap at least 4 ft thick must be placed over the whole area to ensure that no moisture will penetrate even in strong rains or flood-type conditions, or that ground water will not move through the area. As an alternative, cement or fly ash or organic polymers may be considered as soil stabilization agents.

2. *On-Site Storage.*   A landfill may be created directly on site, using a concrete building or structure, or an above-ground cell with an impermeable clay or synthetic liner. A leachate collection system and monitoring wells are also needed.

3. *Transport to a Secure Landfill.*   Excavation of the contaminated soil and removal by truck or rail to an acceptable secure landfill (see page 200) where the soil would be covered and monitored to ensure no leakage of leachate to the environ. The transport mode introduces additional risk into the process, as well as significantly increases the overall cost and exposures of people.

BINGHAMTON STATE OFFICE BUILDING PROBLEM

The concern which has been expressed in many circles where even small amounts of dioxin and dibenzofuran, with other substances, are present, is illustrated by the following report provided by Mr. Richard Bodane, R.A., Chief, Office of Research, Materials and Systems for the State of New York in Albany, dated September 1, 1983.

The Binghamton State Office Building forms a prominent central tower within a governmental building complex shared by the State of New York, the County of Broome, and the City of Binghamton. The State Office Building, completed in the spring of 1973, rises 18 stories (260 ft) above street level and has two subsurface levels. Thirty-three state agencies normally occupy the State Office Building, with approximately 700 employees.

On Thursday, February 5, 1981, the Office of General Services was notified of an electrical fire and subsequent power failure at the Binghamton State Office Building. The fire occurred at approximately 5:35 a.m. when the building was unoccupied with the exception of a security guard and a stationary engineer.

Emergency engineering staff responded, as well as a Department of Transportation oil spills engineer, to find that the fire had been extinguished. Since the markings on each of two tranformers indicated that each contained 1060 gal of

Figure 8.6.   General view of the Binghamton, New York state office building, where transformer-oil containing PCBs in a fire contaminated the whole building. See text for details. Courtesy of G. H. Proper, Jr.

coolant oil containing polychlorinated biphenyls (PCBs), of which 180 gal had leaked out from cracked insulators, the engineering staff initiated necessary action in accordance with the applicable regulations of the Federal Environmental Protection Agency and the New York State Department of Environmental Conservation. New England Pollution Control, a company which maintains a certain expertise in the area of hazardous-material removal, arrived midafternoon and, outfitted with protective suits and respirators, began to clean the floor of the transformer room.

Of the office buildings in the complex, only the State Office Building had suffered damage from the fire, and Office of General Services representatives arranged with county officials to cable electricity from the county building to the state building. The on-site personnel also arranged to borrow a transformer from New York State

Electric and Gas. The transformer would provide heat and electricity so that an efficient clean-up could be accomplished. The next morning personnel continued to establish a temporary electrical system, clean-up of the mechanical-equipment room proceeded, and the Department of Environmental Control was contacted as to the proper disposal of the hazardous materials. A complete inspection of the building revealed that a fine soot had been distributed throughout all areas of the upper floors through the air-conditioning system. The estimated time required for clean-up was extended from several days to several weeks. It was also determined that the New York State Department of Health would arrange to more fully characterize the chemical content of soot and air samples.

Later that day, a meeting was held at which representatives of the city of Binghamton, Broome County, the New York State Departments of Health, Environmental Conservation, Transportation, and Labor, and the Office of General Services were present. During the meeting, an on-site coordinator from the Office of General Services was named, as well as contact persons for all agencies, that were to be involved in the clean-up operation. A Command Post was set up in the Police Bureau of the city building to establish a central communications and coordination center for control of the project.

At the same time, personnel from the Office of General Services were identifying and arranging for alternative office space for agencies displaced from the State Office Building. Space in the Marine Midland Bank building, City Hall, the County Building, and the nearby State Armory was located.

Early Saturday morning, the County Health Commissioner received the test results. The air samples contained 6–62 μg of PCBs per cubic meter, and the soot registered PCB levels of from 10 to 20%. The County Health Commissioner concluded that the air was relatively clean in light of the fact that levels were below the allowable standards promulgated by the Federal Occupational Safety and Health Administration (OSHA). The test results were relayed to Office of General Services in Albany along with samples to be tested further by the State Health Department. All personnel working within the building continued to wear respirators and air-sample tests were repeated on a regular basis.

On Sunday, February 8th, the New England Pollution Control began the clean-up of the overhead areas in the mechanical-equipment room, so that area could again be used to house the power distribution system. It was determined that the soot remaining in the building could best be removed by vacuum and a contractor was retained and started the clean-up of the upper floors on Monday, February 9th. Further power was provided to the building when electricians connected some corridor lighting and one outlet on each floor. Test results from Galson Technical Services indicated that the levels of PCB contamination in the air samples were still well within OSHA standards.

On Monday, arrangements were made for temporary work locations for state employees to conduct business. Thursday, the day of the fire, Friday, and Monday were the only days that state business was hampered due to loss of work space.

Early during the week of the February 9th, the Department of Health staff inspected the building, and arranged for the collection of additional soot and air samples. The team also presented the Office of General Services with a safety plan which was approved and implemented. This plan noted that requirements promulgated by the Occupational Safety and Health Act provided the basic safety program for the operation. The plan addressed security, showers and change areas, safety equipment, respiratory protection, and sampling schemes.

By midweek, the collection of blood samples from all personnel who were working in or had entered the building was initiated. An industrial hygienist from Galson Technical Services was hired to instruct workers and supervisors about proper procedures when handling the contaminated equipment and to advise on the proper use of respirators, protective clothing, and cleaning equipment. The Galson representative studied the safety plan and then observed the cleaning operation to ensure that proper procedures stated in the plan were being implemented.

It was also during that week that the clean-up of the mechanical-equipment room was completed and necessary temporary electrical connections accomplished. The temporary power-distribution system was not ready for installation.

By Friday, February 13th, the Department of Health reported finding PCBs in all soot samples tested, indicating that the contamination was widespread in the building. Based on the findings of the Health Department tests, soot samples from the basement and parking areas were taken. At that time, the New England Pollution Control was asked to expand their professional work force to meet the increased cleaning requirments.

During the week of February 16th, the state decided that, based on the test results from both the Department of Health and Galson Technical Services, the basement parking area would also be treated as a contaminated area. While waiting for the definitive results of the tests of the samples taken from the parking garage, precautionary measures were taken. Revisions to the safety plan were made so that personnel would no longer enter and exit at the parking garage level. By that week, CECOS International, Inc., a licensed waste-disposal firm that had been hired by the state, had removed 230 barrels of toxic waste from the building for deposition in a secure and certified landfill.

Repeat medical examinations were performed on cleaning personnel and others who were determined to have been exposed to any contaminated material.

On Thursday, February 19th, a representative from the National Institute for Occupational Safety and Health reviewed and approved the health and safety plan and trained two persons from the Office of General Services to act as safety observers throughout the clean-up effort.

On the weekend of February 21st, the Department of Health testing confirmed that PCB contamination had been found in samples taken from the parking garage area. The Department of Health recommended that the area be cleaned. On Monday, February 23rd, clean-up of the parking garage area began. After consultation with the Department of Environmental Contamination and the New England Pollution Control, a treatment system for the water discharged from the building was constructed. The Office of General Services made plans for the erection of such an on-site system.

On Wednesday, February 25th, the state health department informed the Office of General Services that their (DOH) analysis firmly indicated that in addition to the previously detected levels of PCBs, the soot also contained lesser amounts of dioxin and dibenzofuran. Upon being informed of this, it was determined that a request for proposal (RFP) to clean the building would be developed and submitted to chemical-pollution specialists. The RFP would be developed by the Office of General Services working in cooperation with the state health department, labor department, and the Department of Environmental Conservation, as well as appropriate federal agencies and acknowledged experts.

On Thursday, February 26th, a press conference was held by the Office of General Services and the State Department of Health at the Empire State Plaza in Albany. At that time, the Commissioner of the Office of General Services announced that clean-up operations were being concluded at their present stage pending the development of the RFP to ensure that not only PCBs were removed, but also tetrachlorodibenzodioxin (TCDD) and tetrachlorodibenzofuran (TCDF). The cleaning of the parking area and the work on the power-distribution systems were permitted to be completed. All other efforts would be concentrated on the removal of debris and would then cease on the evening of Friday, February 27th.

Since the duration of the clean-up effort was not able to be definitively determined, the Office of General Services began considering long-term alternative office space in which to continue the business of government. State agencies had been relocated to temporary quarters elsewhere in Binghamton after the initial 3-day interruption, but the Office of General Services began working with those agencies to secure more permanent quarters to cover the indeterminate period that the state office building would be closed. The vacant Christopher Columbus School was subsequently leased from the Binghamton City School District and would be renovated and made ready for occupancy by the various state agencies. The building would be known as the Binghamton State Office Building Annex.

Also at the request of the Office of General Services, the Department of Health assembled a panel of internationally renowned chemists and physicians, who would meet on April 3, 1981 at New York's Laguardia Airport. The panel would focus on six questions:

What level of human exposure to the various chemicals identified is acceptable?

What kind of testing should be done to monitor the success of any decontamination effort?

Which chemical compounds should be focused on?

Should guidelines be established separately for different routes of exposure (i.e., dermal, ingestion, inhalation)?

Should all areas of the building be subject to the same guidelines regardless of differences in potential human exposure?

What safeguards are necessary to prevent exposure in noncontaminated areas?

Office of General Services personnel continued to remodel, repair, and replace needed heating and plumbing units, and paint and make ready the annex for occupancy. By the week of March 23rd, the first offices were opened at the annex and agencies were advised as to the moving dates. Eventually 20 of the 33 state agencies in Binghamton would be allocated office space at the annex, with the remaining 13 continuing to use office space in the other parts of the city.

After a spot inspection of the garage area by the Broome County Health Commissioner, the Office of General Services was informed that traces of soot were identified on some of the pipes. The Office of General Services instructed New England Pollution Control cleaners to redirect their efforts to the immediate cleaning of the pipes in question. The cleaning was expected to take about 2 weeks since it was being done by hand and each pipe would be tagged as it was cleaned.

By the last week in March, the membership of the expert panel was finalized. Its members included physicians, chemists, toxicologists, epidemiologists, engineers, and representatives from various federal and state agencies, as well as Canada.

On April 1, 1981, the Office of General Services announced that it had hired Versar, Inc., an engineering and research firm to develop a health and safety plan, ventilation plan, and preliminary clean-up and testing plan. Versar recommended that further biological and chemical testing be undertaken. The National Institute for Occupational Safety and Health agreed to act as lead agency in the implementation of the health surveillance and safety plan.

On April 10th, state and local officials formed an Intergovernmental Coordinating Group whose purpose was to bring together, in a formal setting, key state and local personnel to encourage a direct and effective exchange of ideas and information, as well as an updating on the status of the clean-up and testing projects.

In order that preliminary clean-up work could commence, Versar developed a system designed to control the flow of air into the building by drawing it through vents and into a series of filters that would remove pollutants and toxic substances, including dioxins and PCBs, before venting the air to the outside. The Office of General Services began to accept bids for this system, the Air Filtration Pollution Control System (AFPCS), on July 23, 1981. Two weeks later, as part of the health and safety plan, an award was made for construction of an entrance module which would control all passage into and out of the building. The module would be equipped with showers, decontamination areas, and appropriate receptacles for the disposal of contaminated protective clothing.

On August 18th, a public meeting was held in Binghamton to answer citizens' questions regarding any aspect of the state plan or activities relating to the state office building. The health and safety plans and the Air Pollution Control Plan were presented and discussed. A week later, a meeting was held with the editorial board of the local newspapers to inform them of the state activities and to clarify any pertinent matters.

Toward the end of the month of August, the Office of General Services completed an Environmental Impact Assessment in accordance with the State Environmental Quality Review Act. The assessment concluded that the preliminary clean-up of the Binghamton State Office Building would not have any significant adverse effect on the environment. Reasons given in support of that determination were:

Clean-up and ventilation are remedial actions to mitigate existing potential adverse environmental effects. Air exhausting from the filtration and pollution-control system on the roof of the building is intended to be as clean as the existing atmosphere in the Binghamton area.

Analysis of the plan to ventilate the building by the Departments of Health and Environmental Conservation indicated that initial projected air emission dispersion levels will be many times below that approved by the USEPA.

The disposal of solid waste resulting from the preliminary clean-up will be in accordance with approved methods for handling contaminants.

All water used in various segments of the clean-up will be filtered and tested to verify compliance with standards before discharge to the municipal sanitary sewer.

Detailed and intensive sampling methods will be employed to monitor changes in contaminant levels and to ensure worker and community safety. Under the direction of the occupational physician, comprehensive medical examinations will be provided for workers before, during, and after exposure to the contaminants.

The General Health and Safety Plan will govern the conduct of all operations. Standard operating procedures (SOP) will address all safety concerns.

Upon completion of this action, a full assessment of the condition of the building will be made. Based on this assessment, alternatives for further clean-up of the facility will be considered. Any subsequent plan of action will be subject to a separate and full state environmental quality review assessment.

In September, 1981, equipment was moved into the building for tests of the filtration system which was installed on the roof of the building. By the end of the month, the completed entry module arrived and was connected to the building through the corridor which had previously been constructed.

Request for proposal for preliminary clean-up of the building was developed, put to bid, and awarded. Allwash, Inc., of Syracuse, was the successful low bidder. The preliminary clean-up consists of three activities: completing the initial clean-up within the building; removing soot from ceiling panels and surfaces above these panels; and cleaning elevator shafts and mechanical chases.

On October 13th, the Intergovernmental Coordinating Group met in Binghamton with members of the local media. Status reports were presented on the air-filtration pollution-control system, the installation of the entrance module, continued testing, and the medical surveillance plan. A Department of Health physician had begun to hold regular Thursday and Friday office hours in Binghamton for counseling and advising any person having health-related questions or any inquiry concerning the Department of Health's medical surveillance program.

In early December, tests performed on areas of the basement parking garage indicated that the area was usable and parking was allowed in these areas.

The on-going testing of the venting efficiency of the Air Filtration Pollution Control System showed the system worked effectively.

The venting of the building through the Air Filtration Pollution Control System commenced February 1, 1982. The air-conditioning system, which is crucial to the clean-up, has been successfully restored to service. At present, the building temperature is being maintained between 50 and 60°F (10–15.6°C) to allow cleaning personnel dressed in several layers of clothing to effectively proceed with the clean-up.

Since the building is not perfectly square, the ceiling pieces must be returned to their former location. As pieces are cleaned, each is coded in a way that will allow for its reuse. Although this process is a time-consuming one now, it assures the saving of much time and money in the future.

The positive identification of all areas within the building became vitally necessary, since it is crucial that all personnel speak precisely and in common terms when speaking of the building. All offices and spaces within the building have been identified in a systematic and positive manner. The scheme for location identification has resulted in the highest degree of control of information regarding clean-up accomplishments.

Two areas of the clean-up that represent major tasks in their own right are: cleaning of heating/cooling terminal boxes which includes the opening of the box, removal of insulation, cleaning of the box, and its closing. Given the restrictions placed upon clean-up personnel by both the required protective clothing and the various small parts of the terminal boxes, the cleaning of such boxes requires a relatively long time. There are a total of 820 perimeter boxes with another 220 installed in the ceiling.

Second is the cleaning of approximately 285 light fixtures on each of the 18 floors. It requires approximately 90 min to prepare one fixture for cleaning, that is, to take it down and open and remove the lighting element. It then requires over 2 hr to thoroughly clean the fixture and the tracks in the ceiling that hold the fixture. The time consumed equals roughly 5100 fixtures times an average of 4 hr per fixture, or 20,400 person hours for this activity alone. Because these fixtures are metal and therefore very cleanable, and were custom designed for this building, the cleaning rather than replacement of these fixtures is extremely cost effective.

The demand for highest-quality work requires a methodical and meticulous process that cannot be hurried, hence the clean-up requires a relatively long time period, but it will be thorough.

Since the contamination within the building affected the records kept in file cabinets, the State Department of Health advised that it would not be feasible to attempt to clean the documents. Accordingly, the decision has been made to eventually destroy all paper records. However, the Attorney General has requested that we locate, segregate, and temporarily store in the subbasement significant documents relating to pending litigation of state agencies. All documents scheduled for destruction have been shredded or baled and, if of a confidential nature, their destruction by shredding has been confirmed by witnesses. All contaminated materials, in whatever various forms, are being transported to and disposed of at an appropriate disposal site. As of this time, 1200 yd³ of bulk waste and 1377 sealed 55-gal drums have been removed to a high-level waste-disposal site.

Both the 9th and 16th floors of the building have been completely cleaned and serve as sample areas, while also providing personnel with the opportunity to experiment with different testing and cleaning techniques.

Additional activities and accomplishments, which illustrate the volume of work performed inside the building, as of August 27, 1982, included:

Inventory of all equipment and records in building

Removal of all furniture and office equipment to subbasement

Opening of duct shafts to allow for cleaning

Shredding and baling of all paper materials

Removal of all carpets

Removal of all bathroom and corridor accessories

Installation of temporary core lighting on floors 3 through 18

Access and removal of exhaust ducts from restrooms

Preliminary vacuuming of exhaust duct shafts on mens' room side

Removal of secondary duct work in basement

Removal of insulation from terminal boxes in basement

Removal of records and shelving from DOT storage room in basement

Third through 17th floors prepared for vacuuming and washing

Perimeter terminal boxes cleaned on 18th floor

Cleaning of ceiling terminal boxes on 18th floor

Research presently underway by a consultant will aid in determining the level of cleanliness that can be attained for various construction materials, items of equipment, and furniture within the building. This information is being provided, along with risk assessments, to the State Department of Health for the decision as to the appropriate criteria for reoccupancy. The standards that will form the basis for reoccupancy of the building have not yet been finalized.

Even if all the PCBs were removed from all capacitors and transformers, problems would still exist. The restrictions placed on the use of PCBs has resulted in introduction of non-PCB fluids into capacitors and transformers and additional design safety features. The evaluation of non-PCB fluids for fire properties is continuing.

Safety of Personnel

The entry module provides for the safe entry and exit of those who must enter the Binghamton State Office Building and is constructed to include entry facilities, locker areas, showers, restrooms, and security offices. The trailer is located at the basement-level loading dock. The Air Pollution Control System creates a negative pressure throughout the binding and entry module ensuring that the flow of air is from the outside of the building, through the building, and finally filtered through the Air Pollution Control System on the roof.

All personnel entering the building must wear protective clothing and a full-face respirator. The special clothing comprises socks, underwear, sneakers and rubbers, coveralls, an outer "Tyvek" protective suit, and both cotton and rubber gloves. The respirator weighs approximately 4 lb and features both activated carbon and high-efficiency particulate filters. As personnel exit the building through the module, all protective clothing is removed and thorough showers are taken. Respirators are cleaned and their filters are replaced for future use. The outer suit, gloves, and respirator filters are disposed of after each use.

Security

To ensure maximum control of movement into, within, and out of the building, 24-hr daily security is maintained. Not only does the security staff regulate personnel movement, but it also issues and inventories proper safety clothing and equipment. All security systems are tested twice weekly and are monitored by personnel in the central security office.

The fire and smoke alarm system will be automatically activated in the event of any fire or smoke emergency in the building. Five doors in the building are designed as emergency exits and are wired for an alarm to sound should they be tampered with or opened for any reason.

Pressure gauges monitor the operation of the Air Pollution Control Systems. The apparatus is checked daily, and if for any reason the APCS should malfunction, the monitoring devices would automatically notify the security office of the problem and shut the APCS down. In the event of a system malfunction, a magnetic door, located in the entry module, would be released and shut automatically, to prevent any backflow of air from the building to the outside.

Over 265 small bottles and 30 large containers of soot from the building have been collected and delivered to the State Department of Health for use in the testing process. Additional soot samples are being collected on a regular basis from above ceilings and other areas.

The Air Pollution Control Systems are constantly cleaning the building's air as it filters through both carbon and high-efficiency particulate filters for purification before its release to the atmosphere. Periodic testing of the air moving through the system is performed and the system's filters are checked and replaced as required.

Since considerable volumes of waste water are generated by the cleaning procedures, a water-treatment system is in operation at the subbasement level of the parking garage. This system consists of three large plastic tanks, each having a capacity of 13,000 gal. The water is collected in one tank and runs to the second tank through high-rate sand filters that remove large particulates. The water than travels through a series of activated charcoal filters that remove the smallest contaminants. The filtered water is discharged into the sanitary sewer system of the city after it is found to be purified in accordance with the terms of the permit issued by the city of Binghamton.

Restoration and Operation of Building Systems

Electric power was partially restored soon after the fire in February 1981. Since that time, full electrical power with temporary equipment has been restored and is being maintained.

The heating, ventilating, and air-conditioning system (HVAC), which is crucial to the clean-up, has been successfully restored to service. At present, the building temperature is being maintained at between 50 and 60°F (10–15.6°C) to allow cleaning personnel dressed in several layers of nonporous and therefore hot clothing to efficiently proceed with the clean-up.

The elevator system has been restored in phases over the past few months so the system is currently providing service that fulfills the requirements of cleaning personnel. All building systems are operating at levels that are satisfactory to the overall cleaning operation.

Operating Procedures

Maintenance and operations personnel test and ready the building systems each day before entering the building. Five cleaning "teams," each consisting of 11 personnel, including one foreman, then begin their work. Since no personnel are allowed to remain inside the building for longer than 4 hr at one time, each cleaning crew works a shift of approximately 3 hr and 50 min. In addition to cleaning personnel, there are five state inspectors in the building during each shift. They ensure that the work already completed and the work in progress is in accordance with the clean-up standards.

As a means of both basic cleaning and preparing the building for more intensive cleaning, all furniture has been vacuumed and removed from all floors and placed in the subbasement area. The following account of the amounts of furniture and equipment vacuumed and relocated shows the magnitude of the preliminary work:

930 desks
522 tables
850 file cabinets
110 map files
100 lockers
120 racks
200 typewriters
15 postage machines
15 copiers
15 computer terminals
400 miscellaneous items
1950 chairs
325 bookcases
310 storage cabinets
190 stools
52 benches
50 couches
90 recorders
20 postage scales
40 adding machines
5 microfiche readers

Many other smaller items have been compacted and are destined for disposal.

Removal of draperies, carpets, and blank ceiling panels has been accomplished to provide for the most meticulous clean-up. These materials, as well as loose paper and desktop items, have been removed to a secure landfill at Niagara Falls.

As the concluding stages of the primary clean-up operation are begun, we will concurrently begin to engage in those activities and actions which will be necessary to restore the Binghamton State Office Building to a normal operating mode. Activities which will be undertaken can be classed into two major groups: mechanical/operational activities and medical/scientific activities.

The following items are representative to the types of activities which need to be completed to ready the building for reoccupancy from a mechanical/operational perspective:

1. The current primary cleaning program is scheduled to be completed between now and the beginning of August 1983. This activity consists of the washing and vacuuming of all office areas of the building to achieve a high level of cleanliness. Included are the removal of all furniture from the floors as well as the removal of carpeting, draperies, and other appointments. Additionally, lighting fixtures and terminal boxes were removed from their normal stations, completely cleaned, and returned to their proper location. The existing vinyl and ceramic tile flooring will also be removed.

2. As the primary (#1 above) cleaning program draws to conclusion, the building will be divided into basic work areas as follows: upper area—floors 2–18 and lower area—below the second floor. The establishment of these areas will enable workers to more expeditiously function in the upper area while more extensive cleaning activities take place in the lower area.

3. Upon establishing that the air in the upper area meets the preestablished standards for cleanliness, the need to continue air filtration in that area will be unnecessary. This will permit the reconfiguration of the Air Pollution Control System to reestablish normal building ventilation in the upper work area. The Air Pollution Control System units on the roof will be relocated adjacent to the basement area to provide ongoing effective air-pollution control for the lower area of the building as they have for the entire building in the past. At the same time, workers will no longer be required to wear the full range of protective equipment on the upper floors.

4. Once the upper area is established as above described, workers may begin their activities at a regular pace expediting the rate at which building work may continue.

   Service projects to be completed in this phase are related primarily to the restoration of mechanical areas which were opened for cleaning to ensure against the exfiltration of any foreign materials which may be left as a result of previous activities. It is anticipated that all of these restoration activities will take approximately 18–24 months to complete from the date on which they are begun.

5.   While the upper-area projects are being completed as described above, the lower area will receive a thorough cleaning in accordance with the provisions of the Health and Safety Plan which had previously been in existence for the entire facility. Workers will continue to wear protective clothing and utilize the entry module to ensure maximum safety and the Air Pollution Control System will function as in the past. While the workers are active in this phase, they will be cleaning walls, ceilings, and all other surfaces in the lower area. Upon completion of this cleaning phase, a new electrical transformer and related building mechanical and electrical equipment will be installed. Waste material shall concurrently be removed to appropriate landfills. We anticipate that this lower-area activity will take about 18–24 months for completion.

While the ongoing work continues at the building, there will be significant additional activity, on an integrated basis, in the medical/scientific areas related to the ultimate restoration of the facility to use as follows:

1.   Air sampling which was completed late in 1982 for dioxin and dibenzofuran will be repeated with the air-handling systems of the building in operation. Air sampling will be conducted in the normal operating zones of the heating, venting, and air-conditioning system to ensure that comprehensive results are available for review. All results will be transmitted to the Expert Panel for their comments and critique. The first round of samples indicated that the furan levels were bracketing the levels established for reentry and that dioxin was not detectable at the limits which were the capacity of the testing machinery. Repetition of the samples with the air-handling systems in operation and the taking of greater air volumes will allow for analysis and determinations which are representative of the air in the facility as a whole. Results of these tests will guide the state in the final rehabilitation of the facility.

2.   Similarly, the ongoing tests conducted on the operation of the Air Pollution Control System will continue for the entire period during which the Air Pollution Control System is utilized.

3.   We continue the development of a new wipe sample test, to more closely duplicate the actual conditions of everyday work life. Likewise, the continuous monitoring of waste water and the sampling of water utilized in the cleaning and held for purification and disposal will continue throughout the utilization of the waste-water-treatment system.

4.   The Medical Surveillance Program for workers will continue to operate in accordance with the Health and Safety Plan, thereby requiring bimonthly physicals for those entering the lower area of the building and subsequent final and follow-up physical examinations upon completion of a worker's assignment within the building.

5. Further, the monthly industrial hygiene samples which now show the PCB levels below all previously established standards (at 0.2–0.3 $\mu g/m^3$) will be continued as work in the building moves forward.

Financial Data

Through December 31, 1982, restoration of the Binghamton State Office Building has generated expenditures totaling approximately \$8.5 million. This sum reflects payments for the Health and Safety Program; building security; field operations; engineering consultants; investigations and risk analyses; waste-removal planning; the Air Pollution Control System; design fabrication; water treatment and sampling and analyses; testing; purchase and installation of the building entry module and support facilities; installation of special security and alarm systems; boiler repairs; chiller repairs and elevator equipment cooling; emergency electrical work, renovations, and repairs at the Christopher Columbus School; the primary clean-up contract, equipment, materials, and staff; waste disposal; and so on.

The clean-up operation continues and the primary phase is scheduled for completion by early April 1983. The continuing activities beyond that, as herein described, will ultimately take us to restoration of the facility to useful service.

The problems associated with the PCB fire in the Binghamton State Office Building, which we have discussed, are not unique, as was noted at the American Chemical Society national meeting in Washington, D.C. on August 30, 1983. As reported in the *New York Times* on September 1, 1983, page 8, Dr. Arnold Schecter warned that fires in electrical transformers insulated with PCBs were a far greater public health hazard than such widely feared poisons as Agent Orange or the dioxin found at Times Beach. He noted that the toxic soot which results from such incidents are harbingers of an emerging international problem that would get worse unless preventive measures were taken. The Environmental Protection Agency banned further use of PCBs in transformers in 1977, but did not make the ban retroactive. Professor Christoffer Rappe of Sweden noted that the problem was not confined to the United States. Rappe said there have been 10 fires or explosions in electric capacitors containing PCBs over the last 2 years in Sweden, a similar accident in Finland, and at least one more in continental Europe. He estimated there are 1.8 million capacitors and 100,000 transformers containing PCBs in the United States and 200,000 such capacitors and 200 such transformers in Sweden, all posing some risk of releasing their contents, particularly as they age.

Peter Slocum, a spokesman for the New York State Health Department, said a decision on whether to reopen the Binghamton building might be made in a couple of months. The clean-up, as we noted previously, has been underway since February 1981 and is scheduled for completion in summer 1984.

BIBLIOGRAPHY

Ackerman, D. G., *Destruction and Disposal of PCBs by Thermal and Non-Thermal Methods*, Noyes Data Corp., Park Ridge, N.J., 1983.

"Agent Orange Effects: Air Force study fails to Resolve Issue," *Chem. Engr. News*, 4 (March 5, 1984).

Anon., "Fish Stories and Empty Offices," *Time*, April 11, 1983, p. 18.

Bajpai, S. N., and Aquillino, J. W., "Fire Hazards Associated with Non-PCB Dielectrics in Consumer Products, Final Technical Report," CPSC-C-77-0093, Factory Mutual Research Corp., Norwood, MA 02062, FMRC J.I. 1A7N1.RC, RC78-T-44, for U.S. Consumer Product Safety Commission, May 1979, Washington, D.C.

"Bioavailability of TCDD from Soil," *Chem. Engr. News*, 19 (March 5, 1984). (Study by NIEHS suggests TCDD in soil is biologically available in guinea pigs and rats.)

Caglioti, L., *The Two Faces of Chemistry*, MIT Press, Cambridge, Mass., 1983, pp. 28–30, 114, 120, 131, 181.

"Cancer! Cancer!" (Review and Outlook), *Wall Street Journal*, March 14, 1984, p. 30. (Questions the regulatory actions on banning chemicals, such as EDB, on inadequate data).

Council on Scientific Affairs, Advisory Panel on Toxic Substances, "The Health Effects of 'Agent Orange' and Polychlorinated Dioxin Contaminants, Technical Report," October 1, 1981, Department of Environmental, Public, and Occupational Health, American Medical Association, Chicago, Illinois.

Alan Cranston (Democrat—California) introduced bill, S 1651, directing the Veterans Administration to establish a presumption of service connection for certain diseases suffered by veterans exposed to Agent Orange and radiation while in the service. The bill was referred to Veterans' Affairs Committee on August 4, 1983. See *Wall Street Journal*, p. 20, August 12, 1983 Editorial on this bill.

Dangerous Goods Newsletter (monthly), published by Transports Canada, Transport Dangerous Goods Directorate, Transport Canada, Tower B, Place de Ville, Ottawa, Ontario K1A 0N5.

"Dioxin" (Special Issue), *Chem. Eng. News*, 7, 8, 20–64 (June 6, 1983).

Encyclopaedia of Occupational Health and Safety, Third Edition, 1983, Two Vols., 2538 pp., International Labor Office, Washington Branch, 1750 New York Avenue, NW, Washington, D.C. 20006 or Geneva, Switzerland.

"EPA Slow in Controlling PCB's," PB 82-150970, General Accounting Office, Washington, D.C., Springfield, Virginia, 1982.

"EPA Moves to Curb Dioxin Threat," *EPA Journal*, Vol. 10, No. 1, pp. 18–19 (Jan.-Feb. 1984). (Reviews dioxin strategy of EPA to find and control sources of dioxin in the National Dioxin Study).

Evaluation of Thermic Hazards and Prevention of Runaway Chemical Reactions, November 1982, ISBN 0-907-82223-1, Oyez Scientific and Technical Services, Ltd., Bath House, 56 Holborn Viaduct, London EC1A 2 EX (U.K.).

Finch, J. A., Jr., chairman, "Interim Report of the Missouri Dioxin Task Force," submitted to Governor Christopher S. Bond, June 1, 1983; Final report, Oct. 31, 1983. Jefferson City, Missouri.

Garner, W. Y. and Harvey, J., *Chemical and Biological Controls in Forestry*, ACS Symposium Series No. 238, American Chemical Society, Washington, D.C. 1984.

Harris, A., "After the Deluge, a Tainted Town," *Washington Post*, January 6, 1983, p. 1. (Reviews status of the Times Beach potential contamination and possible moving of the town.)

Harris, A., "Etiquette in a Poisoned Town: Do You Accept a Back Rub from a Possibly Contaminated Lady?" *Washington Post*, January 16, 1983, p. B-5.

Harris, A., "Town Struggles with Toxic Legacy," *Washington Post*, January 10, 1983, p. 1 (Times Beach, Missouri story).

Hay, A., *The Chemical Scythe: Lessons of 2,4,5-T and Dioxin*, Plenum, New York, 1982.

Lien, E. J., "Controversies about Toxicities of Dioxins," *Chem. Eng. News* 2 (August 8, 1983).

Long, J., "EPA Outlines Strategy for Tackling Dioxin Problem," *Chem. Eng. News* 21–22 (September 5, 1983). (Plan details how agency hopes to define extent of dioxin contamination, implement clean-up program, and prevent future contamination.)

"Monsanto Finds Dioxin in Soil at Plant," *Chem. Eng. News*, (September 12, 1983). [Two of 26 soil samples from a 10 × 20 ft area along a railroad siding where the herbicide (2,4,5-T was loaded in 1965–1967) showed 49- and 125-ppb dioxin at the soil surface, with one sample, 4–6 in below the surface, higher in dioxin concentration.]

Reggiani, G., "Dioxin, tetrachlorodibenzo-para," *Encyclopaedia of Occupational Health and Safety*, 3rd (rev.) ed., Vol. 1, A–K, International Labour Office, Geneva, Switzerland, 1983, pp. 638–642.

Reinhold, R., "Dioxin Threat Puts U.S. Under Fire," *New York Times*, December 26, 1982, p. 22. (Times Beach, Missouri story, 100-ppb dioxin in some tests.)

Sambeth, J., "What Really Happened at Seveso?" *Chem. Eng.*, 44–47 (May 16, 1983). Vol. 90, no. 10.

"Transcript of Proceedings of Teleconference on Rotating Biological Contractors (RBC's) for Municipal Wastewater Treatment," September 22, 1983, EPA Technology Transfer Teleconference, c/o Dynamac Corp., Dynamac Bldg., 11140 Rockville Pike, Rockville, MD 20852.

Medical Surveillance

Rutstein, D. D., Mullan, R. J., Frazier, T. M., Halperin, W. E., Melius, J. M. and Sestito, J. P., *Sentinel Health Events (Occupational): A Basis for Physician Recognition and Public Health Surveillance*, Amer. Jour. of Public Health, Vol. 73, No. 9, 1054–1062 (September 1983) (Contains list of occupationally related unnecessary diseases, disability and untimely death by condition, occupation, and agent.) See also Chapter 9, this volume.

Papers Presented at the 186th National Meeting of the American Chemical Society, August 29, 1983–September 2, 1983, Washington, D.C. [See detailed abstracts in Meeting Abstracts; note also "Summary: Symposium Updates Health Effects of Dioxins, Benzofurans," *Chem. Eng. News* 26–30 (September 12, 1983).]:

Ayres, D. C., "Destruction of Chlorinated Dioxins and Related Compounds by Ruthenium Tetroxide."

Bell, R. A., and Gara, A., "The Synthesis and Characterization of the Isomers of Polychlorinated Dibenzofurans, Tetra Through Octa." (Describes work to synthesize and identify all 87 PCDF isomers which constitute the homologous series tetra through octachlorodibenzofuran.)

Bergqavist, P. A., "Analysis of Polychlorinated Dibenzofurans and Dioxins in Ecological Samples."

Clement, R. E., "Levels of Chlorinated Organics in a Municipal Incinerator."

Czuczwa, J. M., and Hites, R. A., "Sources and Fates of Dioxins and Dibenzofurans as Told to Us by Sediment Cores."

Davidson, W. R., "The Application of 'Flash' Chromatography—Tandem Mass Spectrometry Analysis of Complex Samples for Chlorinated Dioxins and Dibenzofurans."

Donnelly, J. R., "Quality Assurance Samples for the Dioxin Monitoring Program: Part II: Interlaboratory Study."

Eiceman, G. A., "Adsorption, Chlorination and Photolysis of Selected Chlorinated Dioxins on Fly Ash from Municipal Incinerators Using Laboratory Simulation of Emission Processes."

Elly, C. T., "Federal Procedures Governing Custody, Safety, Transportation, and Costs in the Sampling and Analysis for Dioxins/Furans in the Region V States."

Exner, J. H., "Detoxication of Chlorinated Dioxins." [Various methods of destruction and decontamination using both chemical and physical means (such as ultraviolet light) will be discussed.]

Flicker, M. R., "Evaluation of Veterans for Agent Orange Exposure."

Gierthy, J. F., and Crane, D., "In Vitro Bioassay for Chlorinated Dioxins and Dibenzofurans."

Gross, M. L., "The Analysis of Tetrachlorodibenzodioxins and Tetrachlorodibenzofurans in Chemical Waste and in the Emissions from Its Combustion."

Haile, C. L., "Emissions of Polychlorinated Dibenzo-p-Dioxins and Polychlorinated Dibenzofurans from a Resource Recovery Municipal Refuse Incinerator."

Hay, A., Ashby, J., Styles, J. A., and Elliott, B., "The Mutagenic Properties of 2,3,7,8-Tetrachlorodibenzo-p-Dioxin." (Suggests that TCDD has mutagenic properties and that the chemical does not just act as a promoter by facilitating the growth of a tumor caused by some other agent, but that TCDD may actually initiate the process and cause a tumor to develop.)

Hobson, L. B., "Dioxin in Body Fat and Health Status: A Feasibility Study."

Kaczmar, S. W., Zabik, M. J., and D'Itri, F. M., "Parts per Trillion Residues of 2,3,7,8-Tetrachlorodibenzo-p-dioxin in Michigan Fish."

Kleopfer, R. D., and Kirchner, C., "Quality Assurance Plan for 2,3,7,8-TCDD Monitoring in Missouri."

Kleopfer, R. D., Yue, K., and Bunn, W. W., "Determination of 2,3,7,8-TCDD in Soil."

Kuehl, M. A., Ross, C., and Steiner, C. S., "Bioaccumulation of PCDD's and PCDF's in Fish. Study Design, Methodology and Quality Control."

Lao, R. C., "Analysis of PCDD/PCDF from Various Environmental Sources."

Lawton, R. W., Ross, M. R., Feingold, J., and Brown, J. F. Jr., "Effects of PCB Exposure on Biochemical and Hematological Findings in Capacitor Workers," *Environ. Health Perspectives*, in press, 1984. (Study of 194 workers in G. E. capacitor manufacture.)

Muir, D. C. G., Townsend, B. E., and Webster, G. R. B., "Bioavailability of ^{14}C 1,3,6,8-Tetrachlorodibenzodioxin and ^{14}C Octachlorodibenzodioxin to Aquatic Insects in Sediment and Water."

O'Keefe, P. W., "A Semi-Automated Cleanup Method for Polychlorinated Dibenzo-p-Dioxins and Polychlorinated Dibenzofurans in Environmental Samples."

Rappe, C., "Composition of PCDFs Formed in PCB Fires."

Ryan, J. J., and Williams, D. T., "Analysis of Human Fat for 2,3,7,8-Tetrachlorodibenzodioxin Residues."

Sawyer, C. J., and Stormer, K. A., "Environmental Health, Safety, and Legal Considerations for the Successful Excavation of a Dioxin Contaminated Hazardous Waste Site."

Schecter, A. "The Use of Fat Biopsies to Estimate Patient Exposure to PCDFs, PCBs, and Biphenylenes." (Levels found on a subset of 500 persons established that measurement of fat solubility is direct approach, following the transformer fire containing PCBs at the state office building, Binghamton, N.Y. in 1981.)

Shaub, W. M., and Tsang, W., "Overview of Dioxin Formation in Gas and Solid Phases Under Municipal Incinerator Operations."

Solch, J. G., "Analytical Methodology for Determination of 2,3,7,8-Tetrachlorodibenzo-p-Dioxin in Soils."

Stanley, J. S., and Going, J. E., "Analytical Methods for Measurement of Polychlorinated Dibenzo-p-Dioxins (PCDDs) in Human Adipose Tissues."

Taylor, M. L., Tierman, T. O., Ramalingam, B., and Garrett, J. H., "Synthesis, Isolation, and Characterization of the Tetra Chlorinated Dibenzo-p-Dioxins and Other Related Compounds."

Taylor, P. A., and Slayback, J., "Analysis of TCDD by GC/MS/MS."

Tierman, T. O., "Laboratory Evaluation of the Collection Efficiency of a Stack Sampling Train for Trapping Airborne Tetrachlorodibenzo-p-Dioxin, Tetrachlorodibenzofuran, and Pentachlorophenol Present in Flue Gases."

Webster, G. R. B., Sarna, L. P., and Muir, D. C. G. "K_{ow} of 1,3,6,8-T_4 CDD and O_8 CDD by Reverse Phase HPLC."

Williams, C. H., "Formation of Polychlorinated Dibenzofurans and Other Potentially Toxic Chlorinated Pyrolysis Products in PCB Fires."

Young, A. L., "Environmental Fate of TCDD—Conclusion from Three Long-Term Field Studies."

Young, A. L., "Rationale and Description of the Federally-Sponsored Epidemiologic Research in the United States on the Phenoxy Herbicides and Chlorinated Dioxin Contaminants."

SELECTED REFERENCES

2,4,5-T—Silvex: Hazardous to Human Health?

Almeida, W. F., "Toxicological Aspects of the Herbicides 2,4-D and 2,4,5-T," *Biologico* **40,** 44–51 (1974) (23 refs.).

Anon., "Rebuttable Presumption Against Registration and Continued Registration of Pesticide Products Containing 2,4,5-T," *Fed. Reg.* **43,** 17116–17147 (1978).

Anon., "2,4,5-T, Spina Bifida, and After," *New Zeal. Med. J.* **86,** 99–100 (1977) (3 refs.).

Becroft, D. M. O., and Bakes, M. N., "The Safety of the Herbicide 2,4,5-T," *New Zeal. Med. J.* **86,** 35–36 (1977) (6 refs.).

Bogen, G., "Symptoms in Vietnam Veterans Exposed to Agent Orange [letter]," *J. Am. Med. Assoc.* **242,** 2391 (November 30, 1979).

Boraiko, Allen A., "The Pesticide Dilemma," *Nat. Geog.* **157,** 145–183 (February, 1980).

Colburn, W. A., "A Model for the Dose-Dependent Pharmacokinetics of Chlorophenoxy Acid Herbicides in the Rat: The Effect of Enterohepatic Recycling," *J. Pharmacokinet. Biopharm.* **6,** 417–426 (October 1978).

Cookson, C., "Emergency Ban on 2,4,5-T Herbicide in U.S. [news]," *Nature,* **278,** 108–109 (March 8, 1979).

Courtney, K. Diane, "Prenatal Effects of Herbicides: Evaluation by the Prenatal Development Index," *Arch. Environ. Contam. and Toxicol.* **6,** 33–46 (1977).

Courtney, K. D., Ebron, M. T., and Tucker, A. W., "Distribution of 2,4,5-Trichloro Phenoxyacetic Acid in the Mouse Fetus," *Toxicol. Lett.* **1,** 103–108 (1977) (8 refs.).

Crammer, M., "Toxicology of Families of Chemicals Used as Herbicides in Forestry," in *Symposium on Use of Herbicides in Forestry,* U.S. Department of Agriculture, Washington, D.C., 1978, pp. 53–69 (45 refs.).

Dencker, L., "Tissue Localization of Some Teratogens at Early and Late Gestation Related to Fetal Effects," *Acta Pharmacol. Toxicol.* **39** (Suppl. 1), 1–131 (1976) (214 refs.).

Dmitriyev, V. I., "Harmful Effects of Chemical Substances Used by the U.S. Army in Indochina," *Voenno—Meditsinskii Zh.* **1,** 88–90 (1974) (4 refs.).

Dougherty, R. C., and Piotrowska, K., "Screening by Negative Chemical Ionization Mass Spectrometry for Environmental Contamination with Toxic Residues: Application to Human Urines," *Proc. Natl. Acad. Sci.* **73,** 1777–1781 (1976) (17 refs.).

Dougherty, W. J., Herbst, M., and Coulston, F., "The Non-Teratogenicity of 2,4,5-Trichlorophenoxy Acetic Acid in the Rhesus Monkey (Macaca Mulatta)," *Bull. Environ. Contam. Toxicol.* **13,** 477–482 (1975) (13 refs.).

Ebron, M., and Courtney, K. D., "Difference in 2,4,5,-T Distribution in Fetal Mice and Guinea Pigs," *Toxicol. Appl. Pharmacol.* **37,** 144–145 (1976).

Edwards, W. M., and Glass, B. L., "Methoxychlor and 2,4,5-T in Lysimeter Percolation and Runoff Water," *Bull. Environ. Contam. Toxicol.* **6,** 81–84 (1971).

Galston, A. W., "Herbicides: A Mixed Blessing," *Bioscience* **29,** 85–90 (1979) (52 refs.).

Grant, W. F., "The Genotoxic Effects of 2,4,5-T," *Mut. Res.* **65,** 83–119 (1979) (131 refs.)

Guenthner, T. M., and Mannering, G. J., "Induction of Hepatic Mono-Oxygenase Systems in Fetal and Neonatal Rats with Phenobarbital Polycyclic Hydrocarbons and Other Xenobiotics," *Biochem. Pharmacol.* **26,** 567–575 (1977) (42 refs.).

Gunby, P., "Dispute Over Some Herbicides Rages in Wake of Agent Orange," *J. Am. Med. Assoc.* **241,** 1443–1444 (1979).

Gunby, P., "Plenty of Fuel for Agent Orange Dispute [news]," *J. Am. Med. Assoc.* **17,** 593–597 (1979).

Hardell, L., and Sandstrom, A., "Case-Control Study: Soft-Tissue Sarcomas and Exposure to Phenoxyacetic Acids or Chlorophenols," *Brit. J. Cancer* **39**, 711–717 (1979).

Hartman, Sarah L., "Case History of 2,4,5-T Regulation," *Cong. Rec.* **125**, H 11,394–H 11,395 (November 29, 1979).

Hay, A., "Dioxin Source is 'Safe'," *Nature* **274**, 526 (August 10, 1978).

Hay, A., "Dioxin: The 10-Year Battle That Began with Agent Orange [news]," *Nature* **278**, 108–109 (March 8, 1979).

Hay, A., "Dow Admits 'Poor' Lab Results [news]," *Nature* **278**, 109–110 (March 8, 1979).

Hay, A., "EPA Panel Recommends Postponement of 2,4,5-T Hearings [news]," *Nature* **282**, 124 (November 8, 1979).

Hickling, S., "Pesticides, Health and Ecology," *New Zeal. Med. J.* **79**, 657–659 (1974) (8 refs.).

Hong, S. K., Goldinger, J. M., Song, Y. K., et al., "Effect of SITS on Organic Anion Transport in the Rabbit Kidney Cortical Slice," *Am. J. Phys.* **234**, F 302–307 (April 1978).

Konstantinova, T. K., "Effect of the Herbicide 2,4,5-T Butyl Ester on the Maternal Body and on the Fetus under Experimental Conditions," *Gigiena Sanitariya* **8**, 101–102 (1974).

Koschier, F. J., and Acara, M., "Transport of 2,4,5-Trichlorophenoxyacetate in the Isolated, Perfused Rat Kidney," *J. Pharmacol. Exptl. Therapeut.* **208**, 287–293 (February 1979).

Koschier, F. J., and Berndt, W. O., "In Vitro Uptake of Organic Ions by Renal Cortical Tissue of Rats Treated Acutely with 2,4,5-Trichlorophenoxyacetic Acid," *Toxicol. Appl. Pharmacol.* **35**, 355–364 (1976).

Koschier, F. J., Conway, W. D., and Rennick, B. R., "Renal Transport and Renal Metabolism of 2,4,5-Trichlorophenoxyacetate by the Chicken," *J. Environ. Pathol. Toxicol.* **2**, 927–940 (January–February 1979).

Koschier, F. J., Hong, S. K., and Berndt, W. O., "Serum Protein and Renal Tissue Binding of 2,4,5-Trichlorophenoxyacetic Acid," *Toxicol. Appl. Pharmacol.* **30**, 237–244 (June 1979).

Kutz, F. W., Murphy, R. S., and Strassman, S. C., "Survey of Pesticide Residues and Their Metabolites in Urine from the General Population," in *Pentachlorophenol*, K. R. Rao (ed.), Plenum, New York, 1977, pp. 363–369.

Leng, M. L., "Comparative Metabolism of Phenoxy Herbicides in Animals," in *Fate of Pesticides in Large Animals*, G. W. Ivie and H. W. Dorough (eds.), Academic, New York, 1977, pp. 53–76 (46 refs.).

Lutz, H., "Pesticides and Reproduction in Warm-Blooded Animals," *Bull. Soc. Zoolog. France* **99**, 49–63 (1974) (57 refs.).

Maier-Bode, H., "The 2,4,5-T Question," *Anzeiger Schaedlingskunde Pflanzenschutz* **45**, 2–6 (1972).

National Health and Medical Research Council, "Statements Adopted at 85th Session, Adelaide, June 1978," *Med. J. Austral.* **2**(Suppl. 3), 1–4 (October 21, 1978).

Nelson, C. J., and Holson, J. F., "Statistical Analysis of Teratologic Data: Problems and Advancements," *J. Environ. Pathol. Toxicol.* **2**, 187–199 (September–October 1978).

Nelson, C. J., Holson, J. F., Green, H. G., et al., "Retrospective Study of the Relationship between Agricultural Use of 2,4,5-T and Cleft Palate-Occurrence in Arkansas," *Teratology* **19**, 377–383 (June 1979).

"Ongoing Prospective Cohort Study on Mortality of 2,4-D and 2,4,5-T Among Herbicide Applicators in Finland," Hopes Consulting, Inc., 211 Angus Drive, Aberdeen, MD 21001. (No increase in cancer mortality was detected and distribution of cancer types was unremarkable.)

Sare, W. M., "2,4,5-T and the Problems of Toxicity," *Med. J. Austral.* **1**, 526 (1979) (1 ref.).

Sare, W. M., and Forbes, P. I., "The Herbicides 2,4,5-T and Its Possible Dysmorphogenic Effects," *New Zeal. Med. J.* **588**, 439 (1977) (2 refs.).

Sare, W. M., et al., "Possible Dysmorphogenic Effects of an Agricultural Chemical: 2,4,5-T," *New Zeal. Med. J.* **75**, 37 (January 1972) (4 refs.).

Sauerhoff, M. W., et al., "Fate of Silvex Following Oral Administration to Humans," *J. Toxicol. Environ. Health* **3,** 941–952 (1977).

Scott, R., "Reproductive Hazards," *Job Saf. Health* **6,** 7–13 (1978).

Seiler, J. P., "Phenoxyacids as Inhibitors of Testicular DNA Synthesis in Male Mice," *Bull. Environ. Contamin. Toxicol.* **21,** 89–92 (January 1979).

Seiler, J. P., "The Genetic Toxicology of Phenoxy Acids Other Than 2,4,5-T," *Mut. Res.* **55,** 197–226 (1978).

Simpson, G. R., Higgins, V., Chapman, J., et al., "Exposure of Council and Forestry Workers to 2,4,5-T," *Med. J. Austral.* **2,** 536–537 (November 18, 1978).

Sjoden, P. O., Archer, T., and Soderberg, U., "Effects of 2,4,5-Trichlorophenoxyacetic Acid (2,4,5-T) on Radioiodine Distribution in Rats," *Bull. Environ. Contamin. Toxicol.* **17,** 670–678 (1977) (26 refs.).

Smith, J., "EPA Halts Most Use of Herbicide 2,4,5-T," *Science* **203,** 1090–1091 (March 16, 1979).

Somers, J. D., Moran, E. T. Jr., and Reinhart, B. S., "Hatching Success and Early Performance of Chicks from Eggs Sprayed with 2,4,-D, 2,4,-5-T and Picloran at Various Stages of Embryonic Development," *Bull. Environ. Contamin. Toxicol.* **20,** 289–293 (September 1978).

Somers, J. D., Moran, E. T. Jr., and Reinhart, R. S. "Influence of Hen Dietary Calcium-Phosphorus on the Integrity of the Egg Shell As It Would Influence Hatching Success and the Consequences of Preincubation 2,4,5-T Spraying with and without a High TCDD Level," *Bull. Environ. Contamin. Toxicol.* **19,** 648–654 (June 1978).

Somers, J. D., Moran, E. T. Jr., and Reinhart, B. S. "Reproductive Success of Hens and Cockerels Originating from Eggs Sprayed with 2,4-D, 2,4,5-T and Picloran Followed by Early Performance of Their Progency After a Comparable In Ovo Exposure," *Bull. Environ. Contamin. Toxicol.* **20,** 111–119 (July 1978).

Streisinger, G. "Assessment of Hazards Posed by TCDD," in *Symposium on Use of Herbicides in Forestry,* (U.S. Department of Agriculture, Washington, D.C., 1978, pp. 101–103 (5 refs.).

Sugar, J., Toth, K., Osuka, O., et al., "Role of Pesticides in Hepatocarcinogenesis," *J. Toxicol. Environ. Health* **5,** 183–191 (March–May 1979).

Sugar, J., Toth, K., Somfai-Relle, S., et al., "Research Related to the Herbicide Buvirol, Especially for Possible Carcinogenicity," *IARC Scientif. Publ.* **74,** 167–171 (1979).

Tanimura, T. "Teratogenicity of Chemicals," *Shonika Kijo* (Journal of Pediatrics) **17,** 357–367 (1976).

Toth, K., Somfai-Relle, S., Sugar, J., et al., "Carcinogenicity Testing of Herbicide 2,4,5-Trichlorophenoxy Ethanol Containing Dioxin and of Pure Dioxin in Swiss Mice," *Nature* **278,** 548–549 (April 5, 1979).

Toth, K., Sugar, J., Somfai-Relle, S., et al., "Carcinogenic Bioassay of the Herbicide, 2,4,5-Trichlorophenoxyethanol (TCPE) with Different 2,3,7,8-Tetrachlorodibenzo-p-Dioxin (Dioxin) Content in Swiss Mice," *Prog. Biochem. Pharmacol.* **19,** 177–182 (February 1978).

"2,4,5-Trichlorophenoxyacetic Acid (2,4,5-T) and 2-(2,4,5-Trichlorophenoxy) Propionic Acid (Silvex) (Including All Acid, Ester and Salt Formulations). Emergency Order Suspending Registration for Certain Uses; Suspension Order; Notice of Intent to Cancel Suspended Uses (30 Days to Request Hearing)." *Fed. Reg.* **44,** No. 52, 15,874–15,920 (1979).

"2,3,7,8-Tetrachlorodibenzo-*p*-dioxin(TCDD, 'dioxin')," Current Intelligence Bull. No. 40, NIOSH, Cincinnati, OH [DHHS (NIOSH) Pub. No. 84–104], Jan. 23, 1984 (88 refs.)

"Union Carbide Enters PBC Removal Business," *Chem. & Engr. News,* 17 (Feb. 20, 1984). (New proprietary system will remove PBC from existing transformers.)

"UK Ministry Gives Green Light to 2,4,5-T [news]," *Nature* **278,** 388 (March 29, 1979).

Wagner, S. L., "The Safety of Agricultural Chemicals," *West. J. Med.* **131,** 156 (1979).

Woolfolk, W. and Sanders, R., "Process Safety Relief Valve Testing," *CEP,* Vol. 80, No. 3, 60–64 (March 1984).

9
Medical Care and Surveillance Program for Hazardous Waste Workers

D. Jack Kilian, M.D.
Pamela Harris, M.P.H.

Occupational health programs[1] for employees working around chemicals have been developed by the chemical industry into sophisticated models over the past 40 years. They have evolved into impressive coordinated efforts by management, safety personnel, industrial hygienists, and medical personnel. Some programs have progressed to medical surveillance and epidemiology, far outstripping health efforts in the private sector. The hazardous waste industry, however, is a newcomer with most responsible and well organized companies being less than 10 to 15 years old. It takes time for companies to recognize that these programs are not only good for their employees but also essential for their own survival. This is particularly true for this industry where litigation risks from employees and the community are high. The best legal defense for community challenges is a well documented, healthy workforce where potential exposures are higher.

The chemical industry model program practically always indicates knowledge of the specific chemical involved and the concentration and time of exposure. Furthermore, many times it is possible to search the literature for assistance regarding the human health effect from a particular compound as well as animal toxicology studies. Determination of the hazardous waste environment is very complex and dictates the almost continuous need for maximal personal protective gear. This is difficult to accomplish when the worker does not know exactly what chemical he is working with and assumes that the situation is not critical enough to require the uncomfortable protective equipment. This is the main reason why the hazardous waste industry must put great emphasis on training in the proper use of such equipment and on good medical information about proper first aid after chemical exposure and the human health consequences of acute and chronic exposure to chemical compounds. Such teaching talks are available[2] and are practical teaching tools for small groups of workmen in this industry. Exposure to the training and written rules regarding use of protective equipment must be documented by a signature from the employee. This provides great assistance in enforcing the rules which should have discipline including discharge for noncompliance. Many hazardous waste companies are facing difficult litigation in the negligence and intentional tort areas because the employee maintains, "nobody warned me about the toxicity of the material I was working with" or "nobody really made me wear a respirator." Because of the high rate of contracting certain jobs, the litigation problems are magnified due to third party involvement and lack of workman's compensation protection for the company.

Most chemical hazardous waste workers have a wide variety of jobs to perform and this compounds the safety and personal protective equipment problem. The worker may find himself on a bulldozer excavating a 30 feet deep hole in the ground for the preparation of a secure landfill or operating a backhoe mixing liquid chemicals with fly ash or kiln dust for solidification before transfer to a secure cell, and frequently, to accomplish this, he is the truck driver or front loader operator. Since many waste-disposal sites operate on a 24 hour basis, the worker may work long hours and be responsible for directing a newly arrived vacuum truck to the proper offloading site in the middle of the night. With the wide variety of jobs performed it is impossible at times to know what chemicals he is working around. Sometimes he doesn't know what filter to use in his respirator since the chemical exposure is unknown. Wearing a slicker suit in a hot climate compounds the problem and may produce heat stress. The above worker profile illustrates the great need for even better occupational health programs than the chemical industry now has.

A good occupational health program involves understanding and contributions from an informed management, industrial hygienists, safety engineers, and medical personnel. Preemployment and periodic medical exams should be done as well as, when indicated, biological monitoring. Because of the nature of the chemical waste workers' duties, industrial hygiene monitoring sometimes falls short of assuring a safe workplace. Such monitoring is an essential tool to evaluate an exposure at one point in time and certainly should be done at each waste site. Because of the mixed and frequently changing exposures, figures can be generated which imply a false sense of security from chemical hazards. In many instances biological monitoring (e.g., blood lead, expired air analysis, and urinary porphyrins) not only gives information of exposure but also of accumulation in the human body along with levels that assist in evaluating the human health hazard. With some chemicals, environmental monitoring will yield the most valuable information; with other chemicals biological monitoring is far superior. It takes a well trained expert in occupational health to know which is the most appropriate. Today many hazardous waste companies do not recognize the need for this expertise and are jeopardizing the health of their workmen.

For several years workmen in the chemical industry have had their permissible chemical exposures regulated by federal and sometimes state standards. This was based primarily on the utilization of Threshold Limit Values and Time Weighted Averages of specific chemicals involved. Provisions were set in place for calculating allowable exposures when working more than an eight hour day or when exposed to more than one chemical. It is difficult to apply these principles to a chemical hazardous waste worker because of the problems mentioned previously regarding mixed and frequently changing exposures. In this situation it calls for innovative skills on the part of the industrial hygienist as well as more reliance on periodic physical examinations and biological monitoring. Certain materials used in this industry such as fly ash or kiln dust should be adequately studied for human health hazard, and comprehensive, long range investigations should be established to prevent health effects that, unfortunately, have been rather common in the history of occupational medicine.

9.1. MEDICAL CARE

First aid is defined as the immediate and temporary care given the victim of an accident or sudden illness. The objective is to preserve life, assist recovery, and attempt to decrease or eliminate any serious permanent health effect. The three aspects to any first aid program are:

1. Proper first aid supplies and equipment.
2. First aid training for workers, supervisors, and management.
3. Coordination and education of emergency service providers.

The hazardous waste industry is faced with many specialized first aid considerations. First aid supplies and equipment must reflect the potential for exposure to a large number of possibly unknown chemicals as well as the potential for physical injury. Many factors contribute to the need for extensive first aid training for workers, supervisors, and management personnel. The hazardous waste industry must also recognize the need to educate local emergency service providers concerning the types, extent, and treatments of possible emergency situations which may occur at a site.

A standard first aid kit for any kind of worksite should have a variety of supplies adequate to handle minor first aid treatment on cuts, burns, sprains, and abrasions. In addition, with first aid supplies and equipment for a hazardous waste facility, it is advantageous to include ready access to a stretcher, blankets, splints, and large sterile dressings for burns. First aid equipment should include an easy to operate, portable oxygen unit with airway pieces or a bag mask to assist in the administration of mouth to mouth resuscitation. In many parts of the country snake bite kits may also be needed. Due to the potential for exposure to chemicals, first aid supplies should also include Syrup of Ipecac to induce vomiting, activated charcoal for oral poisonings, and, if the potential for exposure exists, cyanide exposure kits containing amylnitrite and atropine for organophosphate pesticide exposure. Both amylnitrite and atropine require a prescription. It should be noted that amylnitrite has the potential for drug abuse and should therefore be kept under controlled situations.

In situations with the potential for chemical exposure the most valuable first aid supply is a source of large amounts of readily available clean water for irrigation of contaminated eyes and skin. Proper decontamination may be a life saving measure. Emergency showers and eye wash equipment should be placed where chemical exposures are most likely to occur. The water source should be adequate to allow at least a 15 minute shower or eye wash. Decontamination equipment should be clearly identified on a worksite. It should be checked for proper operation daily and should be familiar to everyone on the site.

Special consideration must be given to first aid kits that will be used in the vehicles of hazardous waste transportation workers. It is difficult to include specialized exposure kits, which require prescription drugs and specific administration skills, but a transportation worker should have access to water for chemical decontamination. It may not be possible to carry a water supply large enough for a full 15 minute wash, but any ability to decontaminate is better than allowing the chemical to remain on the skin or in the eye until running water is available. Kits should be inspected routinely and supplies and drugs which are out-dated should be removed. Drugs such as cold remedies containing antihistimines, which may cause drowsiness or affect the safe performance of an individual's job, should not be included in the vehicle first aid kit. These types of drugs should be excluded from any kit used by individuals involved in the operation of any kind of machinery, in a hazardous work situation, or performing any detailed level of work.

The first aid and emergency training of personnel involved in the management of hazardous materials requires planning, special attention, and an ongoing effort. Basic first aid and cardiopulmonary resuscitation training can be obtained from private organizations such as the Red Cross and from some commercial first aid equipment and supply companies. This type of training may not adequately address first aid measures necessary in chemical exposure situations. It may be necessary for a facility to develop its own site-specific first aid training program. The first step in developing a program is to identify and attempt to quantify the chemicals which may present a potential exposure problem. Because of the large number of chemicals in a hazardous waste facility, it may be necessary to group chemicals according to their hazards. Industrial hygiene and environmental monitoring data, material safety data sheets and manifest information, should all be utilized to develop as comprehensive a list as possible. Working with this information, knowledgeable medical professionals can recommend first aid supplies, equipment, and a specific training program.

In general, a first aid training program should include basic first aid measures and cardiopulmonary resuscitation for all workers at the facility. It can be valuable to have an individual at the site trained as a first aid instructor. This makes training an entire workforce less costly and guarantees consistency in training. Beyond the basics, workers need to be trained in the proper first response to chemical exposures. Unlike the chemical, and most other industries, the hazardous waste industry deals with mixtures of often unknown chemicals and has limited control over reactions between chemicals. Speed is of prime importance in the prevention of injury from chemical exposure. It takes valuable time to attempt to identify a chemical or combination of chemicals responsible in an exposure. It also takes valuable time to refer to reference materials to identify the proper first aid measures. In many situations references may not be available. Many worksites handling hazardous waste are intentionally located away from urban areas, placing access to emergency treatment at a time consuming distance. For these reasons it is essential that simple, standardized rules be established for first aid response in chemical exposures. The rules should be set up to cover the worst possible case and every exposure should be treated as such. It is essential that every employee, and particularly supervisors, know these rules and understand that when followed they can save lives, prevent loss of vision or serious permanent injury.

To make training simpler and more effective, exposure situations can be divided by route of exposure. The four routes of exposure are:

1. Inhalation.
2. Skin exposure.
3. Eye exposure.
4. Ingestion.

First aid training in chemical exposures will include the proper use of emergency showers and eye wash equipment, initial burn treatment, administration of mouth to mouth resuscitation and the use of oxygen units. Training should also include information on who to contact for assistance and the order of contact. It can also be very valuable to prepare individuals for the types of questions emergency personnel will be asking. It is important to give emergency personnel as much information concerning the incident as possible to allow them to prepare. If the incident involves a chemical exposure, every effort should be made to identify the chemicals involved after first aid measures have been taken. In remote locations, site directions must be detailed and clear.

The training for supervisory level personnel may require training in specific first aid measures, depending on the presence of specific agents in the workplace. The supervisor must be capable of administrating amylnitrite in cyanide exposures and atropine for organophosphate pesticide exposure.

The training program developed for a site must extend beyond the workforce to the community. Many local emergency service providers are unfamiliar with the types of medical emergencies which may occur at a facility handling hazardous materials. Management of the facility should plan to work with these emergency services. They should be made aware of the physical location and layout of the site, how many employees there are, the type of activities on the site, and the people in charge. Open discussions should be held to determine the kinds of emergencies which may occur, how to handle, treat, transport, and receive injured persons, and what the proper follow-up procedures are. Emergency simulations may be practiced on a routine basis to prepare emergency services as well as site personnel. This kind of relationship with community emergency services benefits the effectiveness of a first aid program and can serve to improve the public attitude regarding the facility.

9.2.   MEDICAL SURVEILLANCE

Medical surveillance is an integrated approach to evaluating the health status of a workforce. A medical surveillance program within industry requires the performance of physical examinations, clinical laboratory studies, a means of storing, retrieving, and manipulating compiled data, and information concerning occupational exposures. It is directed toward the assessment of physical capabilities of workers and the detection of disease or potential disease at the earliest possible stage.

As with the chemical industry, medical surveillance is important to the hazardous waste industry because potential human health effects can not be predicted by industrial hygiene monitoring alone. But medical monitoring as part of a medical surveillance program takes on even greater importance in light of the difficulty of accurately obtaining industrial hygiene data at a hazardous waste facility. No industrial hygiene survey can reflect all of the potential exposure hazards at a site where exposures may change hourly. There are also problems with combined exposures, possible reactions between agents, and the expense of identifying all unknown chemicals in a sample. It must also be remembered that industrial hygiene monitoring measures only the potential exposures. If proper protective equipment is worn, an individual in a high exposure area may have fewer health effects than the individual who is more lax about protecting himself because monitoring has identified his to be a low exposure area.

Industrial hygiene data is a valuable part of a medical surveillance program, but its limitations within the hazardous waste industry must be recognized. The survey data generated through an industrial hygiene program should be used to establish the parameters of the physical examination. Under the Occupational Safety and Health Act (OSHA) special kinds of medical testing and monitoring must be conducted when specific agents are identified in the workplace. These kinds of requirements for the hazardous waste industry can be found in the Occupational Safety and Health Administration Standards For General Industry.[3] Beyond the OSHA requirements, the presence, and therefore potential exposure, of certain agents warrants special medical monitoring because of the possibility of serious health effects and the importance of early detection.

The requirements of the physical examination are determined not only by the OSHA Standards, analysis of industrial hygiene data, potential exposures, and accepted medical practice but also by the results of ongoing medical testing. If the results of an individual or a group indicate a problem, special testing may be required or additional tests may need to be added to the routine medical monitoring of the rest of the workforce. For example, if a routine urinalysis were to reveal albumin, the examining physician may wish to investigate the problem with a series of more sophisticated kidney function tests. If urinary albumin was found during the examination of a number of workers, the kidney function tests may be added as a routine part of every physical examination. Public health considerations may also come into play when designing the specifics of medical examinations. Though it is not the primary function of industry to detect and treat personal injury or illness it is usually to the company's benefit to address many of these health issues.

The hazardous waste industry can not always rely on conventional testing measures. Heavy metal exposure is a good example. Metal exposure in the hazardous waste industry may be from any one or any combination of a large number of metals. The exposure is likely to be sporadic and may be the result of organic or inorganic forms of a metal. Analysis of the blood usually measures the metal which is absorbed and temporarily in circulation before deposition or excretion. Urinary analysis measures the metal which is absorbed but excreted. Both of these conventional methods are exposure-time dependent. With a relatively constant exposure, such as you might find in a heavy metal smelter, blood or urinary measures will probably be an accurate reflection of an individual's exposure. But unless the tests are conducted at the correct time after a sporadic exposure, such as you find in the hazardous waste industry, metal levels measured may fail to detect an exposure problem. An additional consideration is the cost of conducting blood and/or urine analysis on five, six, or even seven different metals.

The hair is a tissue in which trace metals are bioaccumulated. The hair grows at approximately a centimeter a month. Therefore, a three centimeter length of hair analyzed for metals represents an average of three months of exposure. Because of problems in standardizing preparation and analysis techniques and disagreement concerning normal ranges of metals in the hair, hair analysis for heavy metals should be considered only as a screening tool. Hair analysis should not be used as a definite measure of abnormal body burden. If problems are identified through hair analysis they should be investigated further using conventional medical testing methods. The analysis of the hair allows the quantification of several metals often at less than the cost of a single element analysis in the blood.[4,5]

The recognition of the Acquired Immunodeficiency Syndrome (AIDS) and the resulting opportunistic diseases such as cancer and fatal infections has focused great attention on the immune system. Diagnostic tests for evaluation for the immune dysfunction have steadily evolved over the years and are at a stage where objective studies can be made. An excellent example is the immune study done by the Mt. Sinai School of Medicine[6] on 45 Michigan farm residents who ingested food contaminated with polybrominated biphenyls. When the results were compared with 46 nonexposed residents in Wisconsin, there was highly significant evidence of immune suppression when these diagnostic tests were utilized. A study in experimental animals[7] revealed that immune suppression can be produced in mice by doses of TCDD dioxin around 1 microgram per kilogram body weight per week.

This is a minuscule exposure and since this material is easily absorbed through the skin or gastrointestinal tract, concern should be raised for workmen exposed to this material. Extrapolation of animal data to human significance is always difficult, but this work indicates human exposure most certainly should be studied.

For the most part, the specifics of the physical examination must be determined using industrial hygiene data and the past experiences of the chemical industry. There is little data in the literature which addresses the exposure problems or health effect issues in the hazardous waste industry. Medical surveillance programs should include a full preemployment evaluation to assess an individual's capabilities and assure proper placement. As important as establishing the individual's health status, the preemployment examination also serves to document baseline data. This baseline profile is the point from which future health changes will be measured. Periodic examinations are also a part of an effective medical surveillance program. Periodic examinations provide early detection of health effects from exposure to chemical, biologic, or physical hazards. This early detection may prevent progression to a more serious disease state and be a flag to look for early problems in other workers. The time frame between periodic examinations (every six months, annually, once every two years) will depend on the types of work being performed and the kinds of exposures which can be expected. Time frames may also change based on the results of past examinations. An effective medical surveillance program must always be flexible and capable of adjusting to changes.

A great deal of attention should be placed on the development of health history questionnaires used in the hazardous waste industry. The industry has the potential to handle materials which are known carcinogens, mutagens, and teratogens. A health history questionnaire must thoroughly investigate past family health histories and the reproductive history of the worker and spouse. It should contain an exposure history and should attempt to examine lifestyle parameters. Lifestyle parameters such as smoking, alcohol, and drug usage should be quantified whenever possible. The exposure history should contain questions concerning employment in specific industries as well as exposure to specific chemicals. The completed questionnaire should be reviewed by the examining physician before the physical examination, and special tests should be added to the exam if indicated by questionnaire responses.

The data generated from a medical surveillance program must be in useable form. Though most hazardous waste workforces are small it is still difficult to adequately control data through paper files. Computerization of records offers more convenient storage and retrieval and allows the statistical manipulation that would prove to be quite tedious if done by hand. There are several commercial computer software packages which can be purchased, leased, or used through time sharing agreements which handle the kinds of information necessary for a medical surveillance program.[8] Computer programs can also be specifically written for a particular system or company. A computerized database allows something as simple as an automated tick file, which gives the names of individuals who require an updated examination. Reports can be generated which identify and summarize problem areas that may have been missed using paper files. With some types of exposure, OSHA's record keeping requirements are employment plus 30 years.

Computer records are a much more efficient means of storing such bulk. The computerized database also allows correlation between workplace exposures and health status and makes group trend analysis possible. Trend analysis, over time, on one individual can be conducted using paper files without a great deal of effort. Elevation of liver enzymes over the baseline for one worker over the course of employment can frequently provide clues of possible occupational etiology. However, group trend analysis is a useful tool to identify shifts in the normal range and activate remedial changes in the occupational environment before more serious health effects occur. Computerization of the health records and developing programs for periodic analysis of the workforce data can be valuable in identifying an occupational health problem in its early stages. It allows for remedial actions before a serious or permanent health effect can occur. The computerization of medical surveillance information changes file storage to data management.

9.3. SUMMARY AND CONCLUSIONS

In the previous paragraphs we have described the difficult problems that the hazardous waste industry faces to protect the health of its employees. Because of the newness of its management structure there is great difficulty in recognizing the need for sophisticated industrial hygiene and occupational medical programs. Most of the workforce has only a few years of service and the problem of long term chronic diseases has yet to be faced. The turn over rate is high due to the unpleasant work involved and this compounds the problem of assembling a good database for epidemiology study. Legal defenses for alleged health effect are and will be difficult to undertake. Undoubtedly the driving force to improve the occupational health programs will result from the litigation losses cutting sharply into the profit margin. It becomes obvious that those companies who recognize this fact early and take appropriate steps to establish a good occupational health program will survive this highly competitive industry.

We have outlined the steps necessary for an adequate program and indicated which phases are the most important. The cost of health professionals and medical diagnosis is high, but experience in the chemical industry has shown that these costs are essential for survival. It is essential to recognize that the worker health problems in the hazardous waste industry are far more profound.

REFERENCES

1. B.D. Dinman, "Medical Aspects of the Occupational Environment, in Industrial Environment— Its Evaluation and Control." National Institute of Occupational Safety and Health, U.S. Government Printing Office, Washington, D.C., 1973.

2. Medical Graphic Arts, Box 34, Lake Jackson, Texas 77566.

3. "Occupational Safety and Health Administration Standards for General Industry," U.S. Department of Labor, Commerce Clearing House, Inc., Chicago, April 1, 1981.

4. D.W. Jenkins, "Toxic Trace Metals in Mammalian Hair and Nails." EPA-600/4-79-049, U.S. Environmental Protection Agency, Washington, D.C., August, 1979.

5. D.W. Jenkins, "Biological Monitoring of Toxic Trace Metals, Volume I, Biological Monitoring and Surveillance." EPA-600/3-80-089, U.S. Environmental Protection Agency, Washington, D.C., September 1980.

6. J.G. Bekesi et al., "Immunologic Dysfunction among PBB Exposed Michigan Dairy Farmers," *Ann. N.Y. Acad. Sci.,* **320** (1979).

7. R.P. Sharma and P.J. Gehring, "Effects of 2,3,7,8-Tetrachlorodibenzo-p-dioxin (TCDD) on Splenic Lymphocyte Transformation in Mice After Single and Repeated Exposures," *Ann. N.Y. Acad. Sci.,* **320** (1979).

8. "Computer Systems Review," *Ind. Saf. Hyg. News,* 10–14 (August 1983).

SUGGESTED READINGS

1. *Occupational Diseases—A Guide to Their Recognition,* DHEW (NIOSH) Publication No. 77–181, 1978.

2. C. Zenz, ed., *Developments in Occupational Medicine,* Year Book Med., 1980.

3. W. Rom, ed., *Environmental and Occupational Medicine,* Little, Brown, Boston, 1983.

4. *Encyclopedia of Occupational Health and Safety,* 2 vols., 3rd rev. ed., International Labour Organization, Geneva, or U.S.I.L.O. Office, 1750 New York Avenue, Washington, D.C. 20006, 1983.

5. M. Rutter and R. R. Jones, eds., *Lead versus Health: Sources and Effect of Low Level Lead Exposure,* Wiley, Chichester (U.K.), 1983.

6. S. L. Jacobs, *Industrial Hygienists Increases Firms Output and Efficiency,* Wall Street Journal, 33 (March 5, 1984).

7. S. S. Chissick and R. Derricott, eds., *Asbestos: Applications and Hazards,* Vol. 1, 1979; Vol. 2, 1983, Wiley, Chichester (U.K.), 1983.

8. Fawcett and Wood, eds., *Safety and Accident Prevention in Chemical Operations,* 2nd ed., Wiley/Interscience, New York, 1982.

9. M. Key and D. J. Kilian, "Counseling and Cancer Prevention Programs in Industry," in *Cancer Prevention in Clinical Medicine,* G. Newell (ed.), Raven, New York, 1983.

10. W. H. Lederer and R. J. Fensterheim, eds., *Arsenic—Industrial, Biomedical, Environmental Perspectives,* Van Nostrand Reinhold, New York, 1984.

11. C. R. Asfahl, *Industrial Safety & Health Management,* Prentice-Hall, Englewood Cliffs, N.J. 1984.

12. G. Z. Nothstien, *Law of Occupational Safety and Health,* Macmillan, New York, 1984.

13. B. S. Levy and D. H. Wegman, eds., *Occupational Health: Recognizing and Preventing Work-Related Diseases,* Little, Brown, Boston, 1984.

14. S. S. Chissick and R. Derricott, eds., *Occupational Health & Safety Management,* Wiley, Chichester (U.K.), 1984.

15. Dioxin Brief, Missouri Division of Health, St. Louis County Dept. of Community Health and Medical Care, St. Louis, Mo. Jan. 18, 1983.

Appendix 1

National Priorities List by EPA Regions

EPA Region	State	Site Name	City/County	EPA Region	State	Site Name	City/County
01	MA	Industri-Plex 128	Woburn	01	CT	Beacon Hts. Landfill	Beacon Falls
01	MA	Chemical Waste Dump	Ashland	01	CT	Solvents Recovery	Southington
01	MA	Holbrook	Holbrook	01	MA	Hocomonco Pond	Westborough
01	NH	Somersworth Sanitary Landfill	Somersworth	01	NH	—Garage	Londonderry
				01	ME	—Tannery Waste Pits	Saco
01	NH	Route 101	Epping	01	MA	—Chemical	Lowell
01	NH	Sylvester	Nashua	01	MA	—G & H	Woburn
01	ME	—	Gray	01	MA	Groveland Wells	Groveland
01	MA	Acton Plant	Acton	01	RI	—	Lincoln/Cumberland
01	MA	Plymouth Harbor	Plymouth	01	MA	—Engineering	Bridgewater
01	RI	Picillo Farm	Coventry	01	MA	—Resources	Palmer
01	MA	New Bedford	New Bedford	01	NH	Dover Municipal Landfill	Dover
01	CT	Laurel Park, Inc.	Naugatuck Borough				
01	VT	Pine Street Canal	Burlington	01	CT	—Waste Lagoon	Canterbury
01	NH	—	Kingston	01	NH	Auburn Road Landfill	Londonderry
01	RI	—	Burrillville	01	ME	Winthrop Landfill	Winthrop
01	RI	—	North Smithfield	01	VT	Old Springfield Landfill	Springfield
01	MA	—	Dartmouth	01	CT	Old Southington Landfill	Southington
01	RI	—	Smithfield				
01	MA	Charles-George Recycling	Tyngsborough	01	RI	—	North Smithfield
				01	ME	—Salvage Yard	Washburn

01	ME	O'Connor	Augusta	—Rental & Oil	NJ	02	Bridgeport
01	MA	Iron Horse Park	Billerica	Burnt Fly Bog	NJ	02	Marlboro Township
01	CT	Kellogg-Deering Well Field	Norwalk	Old Bethpage Landfill	NY	02	Oyster Bay
01	NH	—Metallurgical	Conway	—Trucking	NJ	02	Fairfield
01	NH	Savage Municipal Water Supply	Milford	—Moreau	NY	02	South Glenns Falls
01	NH	South Municipal Water Supply Well	Petersborough	Brick Township Landfill	NY	02	Brick
01	MA	Sullivan's Ledge	New Bedford	Wide Beach Development	NY	02	Brant
02	NY	Syosset Landfill	Oyster Bay	—Chemical Processing	NJ	02	Carlstadt
02	NY	—Refinery	Wellsville	D'Imperio Property	NJ	02	Hamilton Township
02	NJ	South Brunswick Landfill	South Brunswick	Krysowaty Farm	NJ	02	Hillsborough
02	NJ	—	Pedricktown	—(Chemical Division)	NJ	02	East Rutherford
02	NJ	Ringwood Mines Landfill	Ringwood Borough	Beachwood/Berkley Wells	NJ	02	Berkley Township
02	NJ	Lipari Landfill	Pitman	Vestal Water Supply Well 4-2	NY	02	Vestal
02	NJ	Helen Kramer Landfill	Mantua Township	Frontera Creek	PR	02	Rio Abajo
02	NJ	Price Landfill	Pleasantville	Woodland Rt. 532 Dump	NJ	02	Woodland Township
02	NY	—Abatement Services	Oswego	Hopkins Farm	NJ	02	Plumstead Township
02	NJ	—Industries	Old Bridge Township	Wilson Farm	NJ	02	Plumstead Township
02	NJ	—Landfill	Glouchester Township	Upper Deerfield Township SLF	NJ	02	Upper Deerfield Township
02	NJ	Lone Pine Landfill	Freehold Township	—Sand & Gravel	NY	02	Clayville

National Priorities List by EPA Regions (Continued)

EPA Region	State	Site Name	City/County
02	NY	Love Canal	Niagara Falls
02	NY	—S Area	Niagara Falls
02	NJ	—	Maywood/Rochelle Park
02	NJ	Kin-Buc Landfill	Edison Township
02	NJ	—Chemical	Toms River
02	NJ	—	Bound Brook
02	NY	Batavia Landfill	Batavia
02	NJ	—	Franklin Borough
02	NJ	Lang Property	Pemberton Township
02	NJ	Sharkey Landfill	Parsippany, Troy Hills
02	NJ	Combe Fill North Landfill	Mount Olive Township
02	NY	—Oil Company	Moira
02	NJ	King of Prussia	Winslow Township
02	NJ	—	Elizabeth
02	NY	—	Elmira
02	NJ	Spence Farm	Plumstead Township
02	NY	Port Washington Landfill	Port Washington
02	NJ	Combe Fill South Landfill	Chester Township
02	NJ	JIS Landfill	Jamesburg/South Brunswick
02	NY	Ramapo Landfill	Ramapo
02	NY	—Refining	Colonie
02	NY	Olean Well Field	Olean
02	NJ	Pijak Farm	Plumstead Township
02	NJ	—Resins	South Kearny
02	NJ	Bog Creek Farm	Howell Township
02	NJ	—	Piscataway
02	NJ	Fair Lawn Well Field	Fair Lawn
02	NJ	Rockaway Borough Well Field	Rockaway Township
02	NJ	—	Morganville
02	NJ	Myers Property	Franklin Township
02	NJ	Pepe Field	Boonton
02	NY	—	South Cairo
02	PR	Juncos Landfill	Juncos
02	NJ	—	Newark

02	NJ	—	Kingwood Township
02	NJ	Reich Farms	Pleasant Plains
02	NJ	Monroe Township Landfill	Monroe Township
02	NJ	Florence Land Recontouring	Florence Township
02	NJ	—	Newfield Borough
02	NY	Hudson River PCBs	Hudson River
02	NJ	—	Woodridge Borough
02	NJ	—	Millville
02	NJ	Delilah Road	Egg Harbor Township
02	PR	Barceloneta Landfill	Florida Afuera
02	NJ	—	Florence
02	NJ	Vineland State School	Vineland
02	NJ	Woodland Route 72 Dump	Woodland Township
02	NJ	Landfill & Develop.	Mount Holly
02	NJ	Goose Farm	Plumstead Township
02	NJ	—	Vineland
02	NJ	Williams Property	Swainton
02	NJ	—	Edison Township
02	NJ	—X-ray Corporation	Bayville
02	NJ	—	Gibbstown
02	NY	Niagara County Refuse	Wheatfield
02	NY	Kentucky Avenue Well Field	Horseheads
02	NJ	Asbestos Dump	Millington
02	NJ	Jackson Township Landfill	Jackson Township
02	NJ	Montgomery Township Housing Development	Montgomery Township
02	NJ	Rocky Hill Municipal Well	Rocky Hill Borough
02	NY	Brewster Well Field	Putnam County
02	NY	Vestal Water Supply Well 1-1	Vestal
02	NJ	—	Orange
02	NJ	Sayreville Landfill	Sayreville
02	NY	—Sand & Gravel	Clayville
02	NJ	Evor Phillips Leasing	Old Bridge Township
02	NJ	Mannheim Avenue Dump	Galloway Township
02	NY	Fulton Terminals	Fulton
02	NJ	—Oil & Chemical	Pennsauken
02	NY	—	Lincklaen
02	NY	—Hyde Park	Niagara Falls
02	NJ	Ellis Property	Evesham Township

National Priorities List by EPA Regions (Continued)

EPA Region	State	Site Name	City/County	EPA Region	State	Site Name	City/County
02	NJ	Friedman Property	Upper Freehold Township	02	NJ	Rockaway Township Wells	Rockaway
02	PR	Fibers Public Supply Wells	Jobos	02	NJ	PJP Landfill	Jersey City
02	NJ	—Tank Liners	Bridgeport	03	PA	Moyers Landfill	Eagleville
02	NJ	—Wayne Plant	Wayne Township	03	WV	Leetown Pesticide	Leetown
02	NJ	Ewan Property	Shamong Township	03	PA	Wade (ABM)	Chester
02	NJ	—	Rockaway Township	03	PA	Lackawanna Refuse	Old Forge Borough
02	PR	Vega Alta Public Wells	Vega Alta	03	WV	—Chemical, Inc.	Nitro
02	PR	—	Barceloneta	03	PA	Henderson Road	Upper Merion Township
02	NY	—	Massena				
02	NJ	Tabernacle Drum Dump	Tabernacle Township	03	PA	Mill Creek Dump	Erie
02	NJ	Cooper Road	Voorhees Township	03	DE	Tybouts Corner Landfill	New Castle County
02	NJ	—Delisa Landfill	Asbury Park	03	PA	Bruin Lagoon	Bruin Borough
02	PR	—	Juana Diaz	03	DE	Army Creek Landfill	New Castle County
02	PR	—Del Caribe	Barceloneta	03	PA	—Associates	Macadoo Borough
02	NY	—102nd Street	Niagara Falls	03	PA	Douglassville Disposal	Douglassville
02	NY	—	Cold Springs	03	PA	Osborne Landfill	Grove City
02	NJ	—	Sparta Township	03	WV	West Virginia Ordnance	Point Pleasant
02	NJ	Dover Municipal Well 4	Dover				

Code	State	Site	Location
03	VA	—	Roanoke County
03	PA	Hranica Landfill	Buffalo Township
03	PA	Lindane Dump	Harrison Township
03	PA	Heleva Landfill	North Whitehall Township
03	VA	Chisman Creek	York County
03	PA	Malvern	Malvern
03	DE	—Sand and Gravel Landfill	New Castle County
03	PA	Centre County Kepone	State College Borough
03	PA	—Zinc Pile	Palmerton
03	MD	Sand, Gravel, Stone	Elkton
03	PA	Enterprise Avenue	Philadelphia
03	PA	Presque Isle	Erie
03	PA	Lord-Shope Landfill	Girard Township
03	PA	—Chemical	Lock Haven
03	DE	New Castle Spill	New Castle
03	PA	Industrial Lane Landfill	Williams Township
03	PA	East Mount Zion	Springettsbury Township
03	DE	Old Brime Sludge Landfill	Delaware City
03	PA	Walsh Landfill	Honeybrook Township
03	PA	Berks Sand Pit	Long Swamp Township
03	PA	Taylor Borough Dump	Taylor Borough
03	VA	—	Piney River
03	PA	Old City of York Landfill	Seven Valleys
03	PA	Stanley Kessler	King of Prussia
03	WV	Follansee	Follansee
03	PA	Metal Banks	Philadelphia
03	PA	Westline	Westline
03	PA	Brodhead Creek	Stroudsburg
03	DE	—	Kirkwood
03	DE	Wildcat Landfill	Dover
03	PA	Blosenski Landfill	West Caln Township
03	DE	Delaware City PVC	Delaware City
03	MD	Limestone Road	Cumberland
03	DE	—Steel	New Castle County
03	PA	—	Old Forge Borough
03	VA	Saltville Waste Disposal Ponds	Saltville
03	PA	Kimberton	Kimberton Borough
03	MD	Middletown Road Dump	Annapolis
03	PA	—	Warminster
03	PA	Craig Farm Drum	Parker
03	PA	Voortman Farm	Upper Saucon Township

National Priorities List by EPA Regions (*Continued*)

EPA Region	State	Site Name	City/County	EPA Region	State	Site Name	City/County
03	PA	Resin Disposal	Jefferson Borough	04	FL	Hipps Road Landfill	Duval County
03	PA	Dorney Road Landfill	Upper Macungie Township	04	FL	—	Medley
				04	SC	Geiger	Rantoules
03	WV	Leetown Pesticide	Leetown	04	KY	Distler Farm	Jefferson County
03	PA	Havertown PCP	Haverford	04	TN	Lewisburg Dump	Lewisburg
03	PA	Tysons Dump	Upper Merion Township	04	KY	—	Calvert City
				04	FL	Munisport Landfill	North Miami
03	PA	East Mount Zion	Springettsbury Township	04	SC	—	Fort Lawn
				04	TN	Gallaway Pits	Gallaway
				04	GA	—	Augusta
04	AL	Triana Tennessee River	Limestone/Morgan	04	TN	North Hollywood Dump	Memphis
04	FL	—Schuylkill	Plant City	04	NC	—	Swannanoa
04	FL	—	Tampa	04	MS	Flowood	Flowood
04	FL	—	Pensacola	04	SC	—Nail	Beaufort
04	FL	Davie Landfill	Davie	04	SC	—Chemicals	Beaufort
04	FL	—	Miami	04	AL	—	McIntosch
04	FL	Tri-City Conser.	Temple Terrace	04	GA	—009 Landfill	Brunswick
04	SC	—Wood Preserving	Dixianna	04	SC	—	Florence
04	KY	—	Calvert	04	SC	—	Burton
04	FL	Cabot—	Gainesville	04	SC	—	Rock Hill
04	AL	—	Axis				

04	AL	(Cold Creek)	Bucks
04	GA	Powersville	Peach County
04	GA	—(Areas 1, 2, & 4)	Augusta
04	AL	Perdido Ground Water Contamination	Perdido
04	SC	Scrdi Bluff Road	Columbia
04	KY	A. L. Taylor (Valley of the Drums)	
04	NC	PCB Spills (dumps)	210 miles of roads
04	AL	—	Greenville
04	FL	—Drum Services	Miami
04	FL	—Battery Disposal	Tampa
04	FL	Whitehouse Oil Pits	Whitehouse
04	FL	—Sand Company	Warrington
04	NC	—Sodyeco	Charlotte
04	FL	Zellwood Groundwater Contamination	Zellwood
04	FL	Taylor Road Landfill	Seffner
04	FL	NW 58th Street Landfill	Hialeah
04	FL	Sixty-Second Street Dump	Tampa
04	TN	—(Hardeman County)	Toone
04	FL	—Battery Salvage	Cottondale
04	TN	Murray Ohio Dump	Lawrenceburg
04	FL	—Wood Preserving Company	Whitehouse
04	FL	—	Indiantown
04	FL	—Preserving	Live Oak
04	KY	Distler Brickyard	West Point
04	FL	—Terminal	Ft. Lauderdale
04	FL	Varsol Spill	Miami
04	FL	—	Clermont
04	FL	—	Galloway
04	FL	Pickettville Road Landfill	Jacksonville
04	TN	Amnicola Dump	Chattanooga
04	SC	Scrdi Dixiana	Cayce
04	FL	—Medical Industries	Deland
04	KY	Lee's Lane Landfill	Louisville
04	KY	Newport Dump	Newport
04	FL	Parramore Surplus	Mount Pleasant
05	IN	Parrot Road Dump	New Haven
05	OH	—Chemical	Salem
05	MI	Verona Well Field	Battle Creek
05	MN	Burlington Northern	Brainerd/Baxter
05	IN	—	Zionsville
05	IN	Midco I	Gary
05	OH	Fields Brook	Ashtabula

National Priorities List by EPA Regions (Continued)

EPA Region	State	Site Name	City/County
05	IN	Neal's Landfill	Bloomington
05	MI	Petoskey Municipal Well Field	Petoskey
05	IN	—Waste Oil	Columbia City
05	IL	—Utilities	LaSalle
05	IL	Cross Bros./Pembroke	Pembroke Township
05	MI	—	Wyoming
05	MI	—	Filer City
05	OH	—	Ironton
05	IN	Main Street Well Field	Elkhart
05	MN	Lehillier/Mankato	Lehillier
05	MI	Wash King Laundry	Pleasant Plains
05	OH	Coshocton Landfill	Franklin Township
05	MI	—	Temperance
05	OH	—Oil Company	Jefferson Township
05	OH	Old Mill	Rock Creek
05	WI	Master Disposal Landfill	Brookfield
05	MI	Metamora Landfill	Metamora
05	WI	Municipal Well Field	Eau Claire City
05	WI	Moss-American	Milwaukee
05	MN	—Refinery	Hermantown
05	MN	Joslyn—	Brooklyn Center
05	MN	—	Fridley
05	MI	—Disposal Company	Utica
05	OH	—Iron & Metal	Darke County
05	MN	—	St. Louis Park
05	MN	New Brighton/Arden Hills	New Brighton
05	IN	—Recycling	Seymour
05	MI	—Plating	Cadillac
05	MN	Oakdale Dump	Oakdale
05	IL	—	Greenup
05	MN	—	St. Paul
05	MI	Gratiot County Landfill	St. Louis
05	OH	—	Hamilton
05	IL	—	Waukegan
05	MI	Spiegelberg Landfill	Green Oak Township
05	IL	—Sand & Gravel	Wauconda
05	MI	Ott/Story/Cordova	Dalton Township
05	MI	—Michigan	St. Louis
05	OH	Summit National	Deerfield Township

05	IN	Fisher Calo	Laporte
05	MI	Springfield Township Dump	Davisburg
05	MI	Rose Township Dump	Rose Township
05	MN	—Engineering	Andover
05	OH	Bowers Landfill	Circleville
05	MI	Butterworth #2 Landfill	Grand Rapids
05	MI	G & H Landfill	Utica
05	IL	—Illinois	Marshall
05	MI	Tar Lake	Mancelona Township
05	MI	—	Albion
05	MI	—	Grandville
05	MI	—	Kalamazoo
05	MI	Sparta Landfill	Sparta Township
05	IL	—Solvent/Morristown	Morristown
05	MI	Rasmussen's Dump	Brighton
05	MI	Ionia City Landfill	Ionia
05	IN	—	Lebanon
05	IN	—	Indianapolis
05	OH	New Lyme Landfill	New Lyme
05	MI	—Development Company	Adrian
05	MI	Shiawassee River	Howell
05	IA	—	Council Bluffs
05	OH	Big D Campground	Kingsville
05	WI	Omega Hills North Landfill	Germantown
05	OH	—Lead	Troy
05	WI	Janesville Ash Beds	Janesville
05	OH	Miami County Incinerator	Troy
05	WI	Wheeler Pit	LaPrairie Township
05	MN	—	Cass Lake
05	MN	Bell & Pole	New Brighton
05	WI	Muskego Sanitary Landfill	Muskego
05	WI	Schmalz Dump	Harrison
05	IL	—Sand & Gravel	Wauconda
05	MN	—	St. Louis Park
05	MI	Berlin & Farro	Swartz Creek
05	OH	South Point Plant	South Point
05	MI	—Michigan	St. Louis
05	MI	Littlefield Township Dump	Oden
05	OH	Skinner Landfill	West Chester
05	IL	Belvidere Municipal Landfill	Belvidere
05	IN	Ninth Avenue Dump	Gary
05	MN	—	St. Louis Park

National Priorities List by EPA Regions (Continued)

EPA Region	State	Site Name	City/County	EPA Region	State	Site Name	City/County
05	MI	Southwest Ottawa Landfill	Park Township	05	MI	Duell & Gardner Landfill	Dalton Township
05	OH	Fultz Landfill	Jackson Township	05	OH	E. H. Schilling Landfill	Hamilton Township
05	MI	Forest Waste Products	Otisville	05	MI	Cliff/—Dump	Marquette
05	IN	Lake Sandy JO (Landfill)	Gary	05	MI	Mason County Landfill	Pere Marquette Township
05	IL	—	Waukegan	05	MI	Cemetery Dump	Rose Center
05	MI	—Central	Wyoming Township	05	IL	Byron Salvage Yard	Byron
05	MI	K & L Avenue Landfill	Oshtemo Township	05	MI	Ossineke Ground Water Contamination	Ossineke
05	MI	Charlevoix Municipal Well	Charlevoix	05	MI	—Aviex	Howard Township
05	MI	—Industries	Oscoda	05	MI	—Laundry	Pleasant Plains Township
05	OH	Zanesville Well Field	Zanesville				
05	MI	—Supply Company	Greilickville	05	MI	Verona Well Field	Butler Creek
05	MN	South Andover Site	Andover	05	WI	—Co. Landfill	Sheboygan
05	MI	Kentwood Landfill	Kentwood	05	IN	—Tar & Chemical	Indianapolis
05	IN	Marion (Bragg) Dump	Marion	05	WI	Lauer I San. Landfill	Menomonee Falls
05	OH	—	Reading	05	MN	UNION Scrap	Minneapolis
05	OH	—Reclamation	St. Clairsville	05	WI	Onalaska Municipal Landfill	Onalaska
05	IL	—	Galesburg				
05	MI	—Independent Landfill	Muskegon Heights	05	MN	—Truck & Caster	Fairbault

No.	State	Site	City
05	MI	Sturgis Municipal Wells	Sturgis
05	MN	Wash. Co. Landfill	Lake Elmo
05	MN	—	Minneapolis
05	WI	—Engraving	Sparta
05	MN	—Arsenic Dump	Morris
05	MN	—Arsenic	Perham
05	WI	Delavan Municipal Well #4	Delavan
05	MI	Burrows—	Hartford
05	OH	Powell Road	Dayton
05	IN	Poer Farm	Hancock County
05	WI	City Disposal	Dunn
05	MN	—	Minneapolis
05	WI	—Landfill	Cleveland Township
05	IN	—Chemical	Griffith
05	WI	—Transport	Franklin Township
05	WI	Scrap—	Medford
05	IN	Bennett Quarry	Bloomington
05	WI	Waste Research	Eau Claire
05	MN	St. Louis River	St. Louis County
05	WI	—	Ashippin
06	AR	Crittenden County Landfill	Marion
06	LA	Petro—	Scotlandville
06	TX	Pig Road	New Waverly
06	LA	Cleve Reber	Sorrento
06	AR	—	Jacksonville
06	TX	—	Crosby
06	TX	Motco	La Marque
06	TX	Sikes Disposal Fits	Crosby
06	TX	—	Houston
06	OK	Tar Creek	Ottawa County
06	LA	—Oil Refinery	Darrow
06	NM	South Valley	Albuquerque
06	OK	Hardage/Criner	Criner
06	AR	—Wood Products	Mena
06	AR	Gurley Pit	Edmondsen
06	AR	Frit Industries	Walnut Ridge
06	TX	Highlands Acid Pit	Highlands
06	AR	Cecil Lindsey	Newport
06	TX	—	Grand Prairie
06	MN	—Mining	Milan
06	TX	Harris (Farley Street)	Houston
06	NM	—/Clovis	Clovis
06	NM	—	Church Rock
06	AR	—Control	Ft. Smith
06	LA	Bayou Bonfouca	Slidell
06	TX	—	Bridge City

National Priorities List by EPA Regions (Continued)

EPA Region	State	Site Name	City/County	EPA Region	State	Site Name	City/County
06	TX	—	Conroe	08	CO	Lincoln Park	Canon City
06	OK	—	Tulsa	08	CO	—Wood Products	Denver
06	AR	Cecil Lindsey	Newport	08	CO	Lowry Landfill	Arapahoe County
				08	MT	—Smelter	East Helena
				08	CO	California Gulch	Leadville
				08	SD	Whitewood Creek	Whitewood
07	IA	Labounty Site	Charles City	08	MT	Silver Bow/Deer Lodge	Silver Bow Creek
07	KS	Cherokee County	Cherokee County				
07	MO	Ellisville Site	Ellisville	08	CO	Sand Creek	Commerce City
07	IA	—	Council Bluffs	08	MT	—Smelter	Anaconda
07	KS	Arkansas City Dump	Arkansas City	08	CO	Marshall Landfill	Boulder County
07	KS	—Disposal, Holliday	Johnson County	08	ND	Arsenic Trioxide Site	Southeastern
07	MO	Syntex Facility	Verona	08	UT	Rose Park Sludge Pit	Salt Lake City
07	IA	—	Des Moines	08	CO	Central City, Clear Creek	Idaho Springs
07	MO	Fulbright Landfill	Springfield	08	CO	—	Commerce City
07	MO	Times Beach	Times Beach	08	CO	—Radium Site	Denver
07	MO	Minker/Stout/Romaine Creek	Imperial	08	MT	Milltown Reservoir Sediments	Milltown
07	KS	Johns' Sludge Pond	Wichita	08	WY	Baxter/—Tie Treating	Laramie
07	MO	Shenandoah Stables	Moscow Mills	08	MT	Libby Ground Water Contamination	Libby
07	MO	Quail Run Mobile Manor	Gray Summit				

Code	State	Site	Location
09	AZ	Litchfield Airport Area	Avondale
09	CA	—Wood Preserving	Ukiah
09	CA	—Oil Corporation	Richmond
09	CA	—Oil Sales	Malaga
09	AZ	Indian Bend Wash Area	Scottsdale
09	CA	McColl	Fullerton
09	CA	—Brakes	Coverdale
09	CA	—	Hoopa
09	CA	Jibboom Junkyard	Sacramento
09	CA	—	Selma
09	CA	—Pesticide Storage	Crescent City
09	CA	—Oroville Plant	Oroville
09	CA	Stringfellow Acid Pits	Glen Avon Heights
09	AZ	Tucson International Airport	Tucson
09	CA	Iron Mountain Mine	Redding
09	CA	—	Rancho Cordova
09	AZ	Nineteenth Avenue Landfill	Phoenix
09	TT	PCB Wastes	Pacific Trust Territory
09	AZ	Mountain View Mobile Homes	Globe
09	AS	Taputimu Farm	American Samoa
09	GU	Ordot Landfill	Guam
09	CM	PCB Warehouse	North Marianas
09	CA	—Treating Company	Selma
09	AZ	Indian Bend Wash Area	Scottsdale
09	CA	Coalinga Asbestos	Coalinga
09	CA	Asbestos Mine	Fresno County
09	CA	San Gabriel Area 1	El Monte
09	CA	San Gabriel Area 2	Baldwin Park Area
09	AZ	Kingman Airport Industrial Area	Kingman
09	CA	—	Richmond
09	CA	San Gabriel Valley (Area 3)	Alhambra
09	CA	San Gabriel Valley (Area 4)	La Puente
10	ID	Bunker Hill—	Smelterville
10	OR	Teledyne Wah Chang	Albany
10	WA	Lakewood	Lakewood
10	WA	Com. Bay, Near Shore/ Tide Flat	Pierce County
10	WA	Colbert Landfill	Spokane
10	WA	—(Yakima)	Yakima
10	WA	—	Mead
10	WA	Harbor Island Lead	Seattle
10	OR	—	Portland
10	WA	Pesticide Lab	Yakima

National Priorities List by EPA Regions (Continued)

EPA Region	State	Site Name	City/County	EPA Region	State	Site Name	City/County
10	ID	—	Rathdrum	10	WA	Queen City Farms	Maple Valley
10	WA	—	Kent	10	OR	—Chrome	Corvallis
10	WA	Frontier Hard Chrome	Vancouver	10	WA	Rosch Property	Roy
10	WA	Com. Bay, S. Tacoma Channel	Tacoma	10	WA	American Lake Gardens	Tacoma
10	ID	—	Pocatello	10	WA	Greenacres Landfill	Spokane County
10	WA	Queen City Farms	Maple Valley	10	ID	—	Caldwell
10	ID	—	Pocatello				

NOTE: This list is a dynamic reflection of priorities. Some sites may be cleared in the near future. Names of companies have been deleted in some cases to eliminate misunderstanding. The above list was based on *Fed. Reg.*, 40,658–40,682, (September 8, 1983), which amended Appendix B of *Congress. Fed. Reg.* **40**, Part 300 from the earlier version published in the *Fed. Reg.*, 58,476–58,485 (December 30, 1982). The analyses of the sites are available for inspection in Room S-325, U.S. Environmental Protection Agency, 401 M. Street, SW, Washington D.C. 20460.

Appendix 2

National Priority List Additions Grouped with Similar Hazard Rating System Scores (HRS)

Group 1

EPA Region	State	Site Name	City/County
03	PA	Tysons Dump	Upper Merion Township
08	MT	East Helena Smelter	East Helena
06	TX	—(Fuhrmann)	Houston
02	NJ	—	Vineland
02	NJ	Florence Land Recontouring Landfill	Florence Township
02	NJ	—	Newfield Borough
05	WI	Omega Hills North Landfill	Germantown
05	OH	—Lead	Troy

Group 2

EPA Region	State	Site Name	City/County
05	WI	Janesville Old Landfill	Janesville
04	SC	Independent—	Beaufort
04	SC	—	Beaufort
05	WI	Janesville Ash Beds	Janesville

Group 2 (Continued)

EPA Region	State	Site Name	City/County
05	OH	Miami County Incinerator	Troy
05	WI	Wheeler Pit	LaPrairie Township
02	NY	Hudson River PCBs	Hudson River
01	CT	Old Southington Landfill	Southington
04	MS	Flowood	Flowood

Group 3

EPA Region	State	Site Name	City/County
10	ID	—Railroad	Pocatello
04	AL	—(McIntosh Plant)	McIntosh
05	MN	—	Cass Lake
04	GA	—009 Landfill	Brunswick
05	MN	MacGillis & Gibbs/ Bell & Pole	New Brighton
05	WI	Muskego Sanitary Landfill	Muskego
02	NJ	Ventron/—	Woodridge Borough

Group 3 (Continued)

EPA Region	State	Site Name	City/County
04	SC	—(Florence Plant)	Florence
02	NJ	—	Millville
05	MN	—	Fridley
02	NJ	Delilah Road	Egg Harbor Township
03	PA	Mill Creek Dump	Erie
05	WI	Schmalz Dump	Harrison
08	CO	Lowry Landfill	Arapahoe County

Group 4

EPA Region	State	Site Name	City/County
04	SC	—	Burton
02	NJ	—Tank Liners	Bridgeport
05	WI	—Disposal Landfill	Brookfield
02	NJ	—(Wayne Plant)	Wayne Township
04	SC		Rock Hill
04	AL	—(Cold Creek Plant)	Bucks
04	GA	—(Areas 1,2, & 4)	Augusta
05	OH	South Point Plant	South Point
03	PA	Dorney Road Landfill	Upper Macungie Township
05	IN	Northside Sanitary Landfill	Zionsville

Group 4 (Continued)

EPA Region	State	Site Name	City/County
09	CA	—Asbestos Mine	Fresno County
09	CA	—Asbestos Mine	Coalinga
02	NJ	Ewan Property	Shamong Township
10	ID	—Hide & Fur Recycling	Pocatello
05	MN	—	Brooklyn Center
05	MN	—	Hermantown
05	WI	—	Milwaukee

Group 5

EPA Region	State	Site Name	City/County
01	MA	Iron Horse Park	Billerica
05	WI	—Company Landfill	Sheboygan
05	IN	—	Indianapolis
05	WI	Lauer I Sanitary Landfill	Menomonee Falls
05	MN	—Scrap	Minneapolis
02	NJ	—	Rockaway Township
05	WI	Onalaska Municipal Landfill	Onalaska
05	MN	—	Faribault
02	PR	Vega Alta Public Supply Wells	Vega Alta

National Priority List Additions Grouped with Similar Hazard Rating System Scores (HRS) (Continued)

Group 5 (Continued)

EPA Region	State	Site Name	City/County
05	MI	Sturgis Municipal Wells	Sturgis
05	MN	Washington County Landfill	Lake Elmo
09	CA	San Gabriel Area 1	El Monte
09	CA	San Gabriel Area 2	Baldwin Park Area
06	TX	Pig Road	New Waverly
02	PR	—	Barceloneta
03	PA	Henderson Road	Upper Merion Township
06	LA	—	Scotlandville
03	PA	Industrial Lane Landfill	Williams Township
03	PA	East Mount Zion	Springettsbury Township
02	NY	—Foundry Division	Massena
03	DE	Old Brine Sludge Landfill	Delaware City
05	MN	—	Minneapolis

Group 6

EPA Region	State	Site Name	City/County
01	CT	Kellogg-Deering Well Field	Norwalk
04	AL	—(McIntosh Plant)	McIntosh
04	FL	—	Temple Terrace
05	WI	—	Sparta
01	NH	—	Conway
04	SC	—Wood Preserving	Dixianna
05	MN	Morris Arsenic Dump	Morris
05	NM	Perham Arsenic	Perham
01	NH	Savage Municipal Water Supply	Milford
05	IN	Poer Farm	Hancock County
06	TX	—Creosoting Company	Conroe
05	WI	—Landfill	Dunn
02	NJ	Tabernacle Drum Dump	Tabernacle Township
02	NJ	Cooper Road	Voorhees Township
04	FL	—	Gainesville

Group 7

EPA Region	State	Site Name	City/County
05	MN	—	Minneapolis
09	CA	—Pesticide Storage	Crescent City
02	NJ	—	Kingwood Township
04	GA	—(Augusta Plant)	Augusta
01	NH	South Municipal Water Supply Well	Petersborough
05	WI	Eau Claire Municipal Well Field	Eau Claire City
04	GA	Powersville	Peach County
05	MI	Metamora Landfill	Metamora
02	NJ	—	Newark
02	PR	Fibers Public Supply Wells	Jobos
05	WI	—Disposal Landfill	Cleveland Township
08	CO	Broderick Wood—	Denver
02	NJ	Woodland Route 532 Dump	Woodland Township
05	IN	—Chemical Service	Griffith
05	WI	—Transport & Recycling	Franklin Township
10	WA	Queen City Farms	Maple Valley
05	WI	—Processing	Medford

Group 7 (Continued)

EPA Region	State	Site Name	City/County
02	NJ	Hopkins Farm	Plumstead Township
02	NJ	Wilson Farm	Plumstead Township
06	OK	—Industries	Tulsa
09	CA	—(Oroville Plant)	Oroville
03	PA	Walsh Landfill	Honeybrook Township
02	NJ	Upper Deerfield Township	Upper Deerfield Township

Group 8

EPA Region	State	Site Name	City/County
01	MA	Sullivan's Ledge	New Bedford
05	IN	—Stone Quarry	Bloomington
04	AL	—(LeMoyne Plant)	Axis
04	SC	Geiger	Rantoules
05	WI	—Reclamation	Eau Claire
04	FL	—	Medley
05	MN	St. Louis River	St. Louis County
03	PA	Berks Sand Pit	Longswamp Township
04	FL	Hipps Road Landfill	Duval County
05	WI	—	Ashippin
08	CO	Lincoln Park	Canon City

National Priority List Additions Grouped with Similar Hazard Rating System Scores (HRS) (Continued)

Group 8 (Continued)

EPA Region	State	Site Name	City/County
02	NJ	Woodland Route 72 Dump	Woodland Township
10	OR	—	Corvallis
02	NJ	—	Mount Holly
03	PA	Taylor Borough Dump	Taylor Borough
05	OH	Powell Road Landfill	Dayton
05	MI	Burrows Sanitation	Hartford
10	WA	Rosch Property	Roy

Group 9

EPA Region	State	Site Name	City/County
05	WI	Delavan Municipal Well #4	Delavan
09	CA	San Gabriel Area 3	Alhambra
09	CA	San Gabriel Area 4	LaPuente
10	WA	American Lake Gardens	Tacoma
10	WA	Greenacres Landfill	Spokane County
06	OK	—Petrochemical	Sand Springs
07	MO	Quail Run Mobile Manor	Gray Summit

Source: Appendix—National Priorities List,'' *Fed. Reg.*, 40,678–40,782 (September 8, 1983).
NOTE: For further information contact the Hazardous Site Control Division, Office of Emergency and Remedial Response (WH-548E), U.S. Environmental Protection Agency, 401 M Street, SW, Washington, D.C. 20460. [phone (800) 424-9346 (toll free); in the Washington, D.C. area call 382-3000.] or Hazardous Waste Sites, National Priorities List, HW-7.1, Office of Solid Waste and Emergency Response, U.S. Environmental Protection Agency, August 1983.
To report spills or releases, call 800-424-8802, National Response Center, or (202) 426-2675.

Appendix 3

LEGAL INTEREST IN HAZARDOUS WASTES

A reflection of the increasing interest that the legal profession has in hazardous waste problems, especially as related to land use, long-term health effects, and the impact on the value of real estate, is this scenario for a mock trial. It is taken from the course Land Development Law #21 (fall term, 1983) at the National Law Center, George Washington University, Washington, D.C. It is quoted with permission of Professor James Brown.

CATCHER CHEMICAL CO.—ROPEMAN PARK CITIZENS ASSOCIATION.

Catcher Recycling Co., a subsidiary of the Catcher Chemical Co., Inc., a California corporation, owned, from 1969 to 1980, a parcel of land containing approximately 8 acres, located south and west of the Antelope Valley Country Club and directly west of the subdivision known as Ropeman Park Subdivision of the City of Palmdale, California. Ropeman Park abuts the country club immediately to the south. The Catcher Recycling Co. parcel, roughly triangular in shape, was bounded on the southeast by Amargosa Creek; on the North by Hansa Street and a westward projection due west from the west end of Hansa Street to the point of intersection with the east right of way of Antelope Valley Freeway; thence in a southeasterly direction following the eastern edge of the Antelope Valley Freeway to the point of its intersection with Amargosa Creek. Ropeman Park is directly across Amargosa Creek to the southeast of the Catcher property. Amargosa Creek carries at least a trickle of water during all but exceptionally dry seasons, but other than during rainstorms or occasionally heavy snow melts near its source in the San Gabriel Mountains to the southwest, it seldom carries a flow of more than 300 gallons per minute. On this site, Catcher, in 1973, dug a retention pond covering approximately 200 by 300 feet, or almost 1 1/2 acres. Into this pond, in high water periods, water was diverted by pumping from the creek, to provide sufficient water to fill the recycling of certain of the materials deposited at the site.

The Catcher Recycling Co. provided a disposal site for a number of chemical products, and was used not only by its parent company, Catcher Chemical Co., but by Belgian Girl Paint Co., by Kinkus Stamp Fluid Co., and by Specialties Products Co. On a few occasions, several other firms deposited small quantities of waste chemicals at the site, but their total deposit constitutes less than one percent of the total chemicals identified as being on the site when a first examination was conducted by the USEPA at the instigation of the Ropeman Park Citizens Association, and at the request of the Governor of California.

This first site inspection occurred in July 1981, two months after Catcher Recycling had sold the site to Thornebird Redford. Redford was acting as a "straw" for the City of Los Angeles, which wanted to control this sector of land in order to assure for the proposed Palmdale International Airport a western exit that already had access to the Antelope Valley Freeway. By controlling the land between the limited freeway accesses at the 10th Street Interchange and the "P" Avenue Interchange, PIA would be assured that it could provide a direct route to the Freeway from the east end of the airport, which it felt was a necessity both to avoid further overloading of the "P" Avenue-Division Street route to the major interchange at Palmdale Boulevard, and to avoid routing all airport traffic through slow-moving streets in downtown Palmdale. This access would also relieve much of the Air Force Plant No. 42 traffic from the other airport access routes and channel most of the substantial Lancaster traffic directly out of the area. While T. Redford still holds technical legal title to this Catcher parcel as of September 1, 1981, she bought it under authorization from the City of Los Angeles and its Department of Airports. She took title in trust for them, to deliver it upon demand, and upon receipt of purchase and transactional and holding costs plus a commission of 7% of the agreed purchase price, which she negotiated.

On the site at the time of the sale were three bulk storage tanks, a large quantity of metal drums and barrels, and some smaller plastic containers, plus a number of 5-gallon metal tins containing off-specification paint from Belgian Girl Paint Co.

The residents of Ropeman Park had been complaining to the City of Palmdale, to the County, to the State, and to Catcher Recycling Co. for a period commencing in 1978 when, after a heavy spring rain which produced some flooding along the Creek, strong and objectional fumes or odors were persistent for several days and a substantial fish kill was evidenced. This same odor was noticed in the drinking water, which for this sector of the community is derived from deep wells. After notification, the City of Palmdale conducted some tests and cautioned property owners serviced by the Quartz Hill Water District, whose wells tapped the aquifer supplying Ropeman Park, other nearby Subdivisions, and the Country Club, that they could be drinking polluted water when the odor was noticed. When the condition persisted and continued to worsen, the Country Club closed its swimming pool. In the spring of 1980, when unusually heavy rains occurred on several different occasions, three workers in the Northwest Sewage Treatment Plant were overcome by toxic fumes. This plant, just off the Sierra Highway on Avenue M-8, also serves Plant No. 42. It is connected to a combination store-sanitary system which subjects the plant to peak loading during periods of heavy run-off. After this episode, officials of the State Department of Waste Resources conducted an examination and concluded that leachate from the Catcher site runs into storm sewers which discharge into the sanitary sewer system and is thus carried to the treatment plant. State inspectors also determined, during this inspection, that one of the three bulk storage tanks on the site was contaminated with PCBs. They further determined that Catcher itself did, or had allowed some other entity on occasion to dump recycling residue directly into the pond on the site. In addition to the PCBs, the state inspectors also found dioxin (from 2,4,5-trichlorophenol) on the site.

In addition to the companies listed above, the state inspectors found that the federal government had deposited small quantities of dangerous chemicals on this site from time to time, which most probably originated either at Air Force Plant No. 42, or at Edwards Air Force Base. Thoroughly alarmed by this time, Ropeman Park residents, whose children used to build paddle wheels in Amargosa Creek and go wading and fishing in it (ponds upstream in the creek are intermittently stocked by the state fish and game department), began to organize to try to force some solution to what they now realized was a severe threat to their health and that of their children. At about this time they also began to notice that on occasion lids from some drums stored on the site would "pop" and literally fly up into the air. Since that time, there have been intermittent fires at the site. Firemen from the station on "M" Street, between 3rd Street, West, and Division Street, in whose zone Ropeman Park falls, said that they cannot expect to get to the site in less than seven minutes after receiving notification. They have concluded that the fires start by chemical reaction and by spontaneous combustion, and that fumes from some combinations of chemicals on the site are highly toxic. The chemicals on the site were given a detailed examination and identification and their quantities tallied when the USEPA inspection team first surveyed the site in September 1981. This tally is available as a separate document. The parties have stipulated that there were three chemicals dumped by Catcher: dioxin, formaldehyde solutions, and trichlorethylene. Among the several isomers of dioxin present the most prevalent was 2,3,7,8-tetrachlorodibenzo-p-dioxin (TCDD) one of the most potent toxins known to science, in the opinion of some persons. It is acutely toxic at low doses; it causes cancer, birth defects, mutations, and fetal death in laboratory animals. Dioxin is persistent in the environment and is bioaccumulated (bioaccumulation factor, 6,000 in fish). The official USEPA Water Quality Criterion for 2,3,7,8-TCDD for protection of human health is zero. The level of exposure to dioxin which can be expected to pose a cancer rate of one additional cancer case per million persons is 0.000000046 micrograms per liter (parts per billion or ppb). TCDD has been detected in Amargosa Creek at 220, 80, and 18 micrograms per kilogram (ppb).

Similar human health limits and other data regarding some of the other listed chemicals are included in the files we have received from Bill Walsh and John Wheeler at USEPA, and will be made available to the interested parties.

This game's interactions will be initiated by the Ropeman Park Citizen's Association and by USEPA, directed at the Catcher Chemical Co. Both Catcher Chemical and its subsidiary, Catcher Recycling Co., will be represented by the two class members designated as Catcher Chemical Co. If the State of California becomes involved, it will be represented by Canara, and assisted by the Governor et al. The United States, as a defendant, will be represented by the President et. al. Palmdale's Mayor/City Attorney will represent that city, and the DOD/President will represent the Air Force. Los Angeles will be represented by its Mayor. Appropriate assistance will be rendered as needed by the other designated governmental offices. The other named Users of the waste dump site will be assigned when and if one of the principles above decides that any of them need to be specifically represented in the game, and so advises the Judge.

The initiatives, as the game moves into the September-December 1981 game time period, rests with USEPA and with Ropeman Park Citizens Association. The Federal establishment will publicize former Secretary Watt's recent announcement that his intention is to site hazardous waste facilities on government property. However, we will accelerate that pronouncement, making it first during the January-April 1982 period. It is suggested that Ropeman select several (not over four) representative plaintiffs.

Index

The widespread media interest in hazardous and toxic chemicals—especially the extensive coverage of improper or inadequate disposal practices—has seriously damaged the public image of the chemical and engineering profession and caused a considerable financial drain on the industry. The term "toxic chemical" has almost become one word.

This book offers a balanced, unbiased view of the latest scientific information about hazardous and toxic materials, their containment, and their availability to man, animals and plants. First, *Hazardous and Toxic Materials* takes a close look at the laboratory where most of these materials originate and then proceeds into fire and explosion hazards and their detection. Avoiding excessive "horror stories," it details the personal protection for the body, head, feet, and respiratory system that's necessary in the waste site investigation and clean-up environments. The legal applications of RCRA and SUPERFUND—the enforcement laws and financial base—are discussed at length. There's also an updated analysis of the most widely misunderstood chemical of all time, dioxin. A special Appendix lists 546 "National Priority Sites" which are broken down geographically.

Health and environmental officials, local action groups, chemists, engineers, technical advisors, laboratory workers, students, and anyone concerned with this topical issue will find here the scientific facts behind the newspaper headlines. Assuming a positive, problem-solving approach, *Hazardous and Toxic Materials* emphasizes the importance of alternative disposal methods and shows how to control and prevent future environmental disasters.